Contents

Preface

The importance of quality assurance has become widely appreciated during the last decade and procedures to ensure product quality have now been implemented by almost all large companies. The adoption of rigorous quality control systems is not yet so widespread among small- to medium-sized companies, but international market forces and the trend towards imposition of contractual quality standards by large companies on sub-contractors means that such smaller companies will have to fall in line rapidly. Increasingly competitive world markets mean that no company will be able to survive for very long if it does not achieve the necessary quality assurance targets.

The British Standards Institution, the International Standards Organisation and the European Committee for Standardisation have agreed on a common framework of procedures for implementing quality control and providing assurance about the quality of products. This means that, if the published guidelines are followed, implemented quality control systems will have international significance and no problems should be encountered in exporting products across national boundaries.

Quality assurance has an interdisciplinary nature and, if it is to be achieved, bridges must be built between all functions in a company and especially between engineering and management functions. One major reason for the success of Japanese industry in achieving a high-quality record for its products has been attributed to the creation of quality circles, which involve regular meetings of representatives of all functions in a company and ensure that quality control is a corporate, cohesive policy which permeates through the entire company instead of being a fragmented policy which operates effectively in some departments only. Whether achieved by formal quality circles or otherwise, an essential component of quality assurance is full management commitment to it and the provision of adequate funding for the creation and maintenance of quality control systems. A study of the Japanese success in achieving high product quality has also shown that a further important ingredient is the provision of training programmes in quality assurance at an appropriate level for all company personnel. Such training ensures that personnel fully appreciate the importance of quality assurance and understand the contribution that they are required to make towards achieving it.

Measurement and calibration procedures are an essential component within quality control systems, both for monitoring the values of quality-related process parameters at various stages of production and also for inspecting and testing the final product. Measurements at intermediate and final stages of production provide the means for assessing the degree of success of the quality control systems operated. Such measurements are the means by which the customer is assured of the quality of the product. This requires in turn that the quality and accuracy of the measurements must also be guaranteed by a properly managed system of instrument calibration.

Previously, quality control system management and quality measurement techniques have been treated as separate subjects and described in separate texts. The unique nature of this present text is that it provides an integrated approach to these two interdependent subjects within a single text. Besides avoiding the need to consult separate texts, there is a significant benefit in having such an integrated approach. It enables measurement and calibration procedures to be considered in the proper context with respect to their importance in the operation of quality control systems and the attainment of quality assurance targets.

The method of approach has been to present the subject of quality assurance first, highlighting its importance and describing the procedures involved. This is followed by a discussion of the general principles of the measurement and calibration procedures necessary for the operation of quality control systems, and a description of the mechanisms for assessing the sources of measurement error and quantifying their effect. Finally, a series of separate chapters is devoted to each of the commonly measured quality-related quantities. Each of these chapters reviews the appropriate measuring instruments available and discusses the relevant calibration procedures.

As stated earlier, quality assurance is a company-wide activity to which the company's management must be fully committed. The role of management is to assign responsibility for meeting quality targets, to delegate authority for operating quality control systems and to create accountability for the operation of all aspects of a quality plan. To fulfil this role, management must have an awareness and understanding of all activities relevant to quality assurance. To facilitate this, the technical level of this text has been deliberately kept as low as possible. Where necessary technical detail has been provided for the benefit of staff implementing quality control and measurement/calibration procedures, the aim has been to provide explanations in simple terms for the benefit of the non-technical reader.

1

Introduction

In this age of high-intensity, international competition in the market place, assurance to customers of high standards in quality control has become vitally important to all manufacturers and providers of services. Such quality control procedures must be orientated towards preventing quality problems occurring, rather than being mere fault-detection systems which allow faults to be put right before a customer complains about them. When introducing quality control procedures, it is important that employees at all levels in a company are aware of the reason for them, understand fully how to operate them, and co-operate enthusiastically in implementing them. While it is sensible to appoint a quality control manager to design and monitor quality control systems, the responsibility for quality control must never be seen as the responsibility of this one person alone. All company personnel must be encouraged to share in the duty of maintaining good quality and take pride in it.

It is also important that such procedures should evolve and develop over a period of time. They must not stay the same for ever after instigation. Rather, regular review is necessary to ensure that quality control procedures continue to be efficient and remain appropriate with technological development. Such reviews must also take changes in market forces into account and fully monitor customer satisfaction, need and expectations.

Quality circles are a useful method of attaining these ideals. A quality circle (see Figure 1.1) consists of a group of people who collectively represent all those functions within a manufacturing company that can have an effect on quality. For instance, such a group must represent goods packing and delivery sections as well as shop-floor operatives from the production departments. The discussion which takes place at periodic meetings of the quality circle fulfils several functions. Firstly, it ensures that thoughts about quality maintain a high profile throughout the company. Secondly, it provides a feedback mechanism whereby breakdowns in, or difficulties with, a quality system can be reported

1

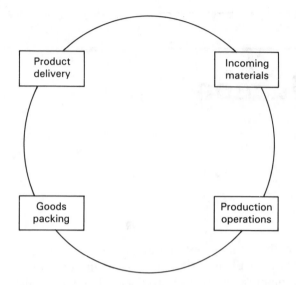

Figure 1.1 The quality circle

and suggestions for improvement made. Thirdly, by giving everyone such a personal involvement in achieving high quality, an atmosphere is generated where people take pride in their work, understand the reasons for the quality control procedures implemented and are fully committed to their operation.

All industrialized nations publish standard procedures for quality assurance and the maintenance of calibration equipment to ensure the accuracy of measurements used in quality checks. This book is primarily concerned with that aspect of quality assurance which is concerned with the calibration of measuring equipment, although the wider aspects of quality control are considered in Chapter 2.

1.1 Quality standards

In the United Kingdom, the appropriate procedures for attaining quality assurance are defined in document *BS 5750* (Parts 0–4).[1] This was first published by the British Standards Institution in 1979 and, since that time, has been adopted in a wide range of industries. This first version has been modified in collaboration with the International Standards Organisation in the light of user experience. A revised version was published in 1987 by both the British Standards Institution[1] and the International Standards Organisation[2] as two separate but identically worded documents (ISO versions are numbered *ISO 9000– ISO 9004*). At the end of 1987, the procedures were also adopted by the European Committee for Standardisation and published as identically-worded documents numbered *EN 29000–EN 29004*.[3] Prior to 1987, a separate document, *BS 5781*, existed which detailed the necessary measurement and calibration procedures associated with quality assurance systems, but this became obsolete when these procedures were subsumed within *BS 5750* in 1987.

The contents of the various documents detailing quality standards are as follows:

BS 5750 part 0 section 0.1: Quality systems – guide to selection and use (identical standards: ISO 9000/EN 29000)

This gives general guidance on quality management. It also advises on which quality assurance model out of the three contained in *BS 5750* parts 1, 2 and 3 (*ISO 9001, 9002, 9003/EN 29001, 29002, 29003*) to use. Each of these alternative models provides a structure for a contractual quality agreement between supplier and purchaser. Together, the models describe three different forms of functional or organizational capability which are suitable for two-party contractual purposes. Guidance is also given about how to modify and tailor these models for special contractual situations. The guidance given about the choice of models can be summarized as follows:

- *BS 5750* part 1 (*ISO 9001/EN 29001*) should be used when the contract specifically requires design effort and the product requirements are stated principally in performance terms. In this case, the supplier's capabilities in the functions of design, production, installation and servicing must be demonstrated in order to provide confidence in the performance of the product.
- *BS 5750* part 2 (*ISO 9002/EN 29002*) should be used when the specific requirements for a product are stated in terms of an established design. In that case, confidence in the product performance can be won provided that the supplier's capabilities in production, inspection/test and installation are demonstrated.
- *BS 5750* part 3 (*ISO 9003/EN 29003*) should be used when the conformance of a product to contractual specifications can be demonstrated by the supplier's capabilities in inspection and test functions only.

BS 5750 part 0 section 0.2: guide to quality management and quality system elements (identical standards: ISO 9004/EN 29004)

This provides more specific guidance on quality management. It lays down the general requirements of quality system management, documentation, audits, costing, training and the control of measurement and testing equipment.

BS 5750 part 1: Quality systems – model for quality assurance in design/development, production, installation and servicing (identical standards: ISO 9001/EN 29001)

This contractual model is for use when conformance to some specified requirements is to be demonstrated by the supplier of goods over a range of functions which include design–development, production, inspection–test, installation and servicing.

Control of the design process is particularly important in this model and the supplier is required to establish and maintain procedures which control the product design and ensure that specified requirements are met. Particular duties imposed are to identify clearly the person responsible for design control, to use qualified personnel in the design team and to provide the team with adequate resources. All design procedures have to be documented and all input and output associated with the design process must also be recorded. Mechanisms for changing design procedures in response to changing product requirements must also be documented.

Production processes must be carried out under controlled conditions which guarantee the quality of the product. Similar control must be ensured for the quality of raw materials and part-finished components bought in from sub-contractors.

Any product installation procedures which can affect quality must be rigidly controlled. Handling, storage, packaging and delivery functions must also be designed to avoid damage to products.

To demonstrate proper control of inspection and test functions, the product supplier must establish and maintain a system of records which show that the product has passed inspections and/or functional tests.

Equipment used to make quality-related measurements at any stage of production, inspection or test must be controlled such that measurement uncertainty is known. Calibration must be carried out at defined intervals of time.

Proper documentation must be maintained of all aspects of the quality assurance system.

BS 5750 part 2: Quality systems – model for quality assurance in production and installation (identical standards: ISO 9002/EN 29002)

This contractual model is for use when conformance to some specified requirements is to be demonstrated by the supplier of goods over a range of functions including production, inspection–test, installation and servicing, but the design–development function is specifically excluded. The requirements are the same as specified in *BS 5750 part 1* apart from the absence of design control.

BS 5750 part 3: Quality systems – model for quality assurance in final inspection and test (identical standards: ISO 9003/EN 29003)

This contractual model is for use when conformance to some specified requirements is to be demonstrated by the supplier of goods only in final inspection–test functions and delivery of the product. The procedure for implementation is described in those parts of *BS 5750 part 1* which cover inspection–test functions and the control of handling, storage, packaging and delivery.

BS 5750 part 4

This part of *BS 5750* give guidance on interpreting the requirements specified in *BS 5750 parts 1–3*. There is no corresponding ISO or EN document.

1.2 Summary of quality requirements

BS 5750 describes a system of standard procedures for the assurance of quality in design, manufacture and installation. These procedures apply equally to the manufacture of goods and to the provision of services. The major requirements specified by *BS 5750* are summarized below:

1. The supplier shall establish and maintain an effective, economical and demonstrable system to ensure that material or services conform to the specified requirements. The system must demonstrate adequate control over the quality of incoming raw materials and components bought in from suppliers external to the company.

2. The responsibility, authority and interrelation of all personnel who manage, perform or verify work affecting quality shall be defined. One designated person shall be given overall responsibility for ensuring that the requirements of *BS 5750* are met.

3. All personnel involved in quality control procedures must be adequately trained.

4. The quality system established shall be periodically and systematically reviewed to ensure its continued effectiveness. Records of the review shall be maintained. Any statistical techniques used to sample products for quality at any stage in the production process between input and output must also be explained.

5. The supplier shall provide, control, calibrate and maintain inspection, measuring and test equipment suitable to demonstrate the performance of the product to the specified requirements. Equipment shall be used in a manner which ensures that measurement uncertainty is known.

6. Measurement and calibration for this equipment shall be in accordance with the procedures set out in Section 1.3.

7. The inspection–test status of products shall be identified by a documented system of markings. A proper system must be in operation which deals effectively with products that fail quality tests. A procedure for responding to such failures by reviewing the effectiveness of quality control procedures must also be documented.

8. The operation and maintenance of all quality control procedures shall be fully documented. This documentation will typically take the form of a quality manual. Such quality control procedures must extend beyond the production area to handling, storage, packaging and delivery functions.

9. Copies of the quality manual, or appropriate parts of it, shall be available at all locations where operations essential to the effective functioning of the quality system are carried out.

10. All documentation must be dated, have its revision number identified, and be maintained in a clean and legible condition.

11. A documented system must be established whereby the quality manual is modified or has additions made to it to keep it up to date. The responsibility for this should be clearly specified and must normally be the person who originally produced the quality manual.

12. The quality management system must contain documented provisions for disposing of obsolete documentation.

1.3 Summary of measurement and calibration requirements

BS 5750 lays down procedures to be followed when selecting, using, calibrating, controlling and maintaining measurement standards and measuring equipment. A summary of the requirements which it imposes is given below:

1. The supplier shall establish and maintain an effective system for the control and calibration of measurement standards and measuring equipment. *Note*: complete in-plant calibration is *not* essential if these services are obtained from sources that comply with the requirements of this standard.
2. All personnel performing calibration functions shall have adequate training.
3. The calibration system shall be periodically and systematically reviewed to ensure its continued effectiveness.
4. All measurements, whether for purposes of calibration or product assessment, shall take into account all the errors and uncertainties in the measurement process.
5. Calibration procedures shall be documented.
6. Objective evidence that the measurement system is effective shall be readily available to customers.
7. Calibration shall be performed by equipment traceable to national standards.
8. A separate calibration record shall be kept for each measuring instrument. These records must demonstrate that all measuring instruments used are capable of performing measurements within the designated limits. The record for each instrument shall contain *as a minimum*:
   - a description of the instrument and a unique identifier;
   - the calibration date;
   - the calibration results;
   - the calibration interval (plus date when next calibration due).
   Some or all of the following information is also required in the calibration record, according to the type of instrument involved:
   - the calibration procedure;
   - the permissible error limits;
   - a statement of the cumulative effects of uncertainties in calibration data;
   - the environmental conditions required for calibration;
   - the source of calibration used to establish traceability;

- details of any repairs or modifications which might affect the calibration status;
- any use limitations of the instrument.

9. All equipment shall be labelled to show its calibration status and any usage limitations (if practicable).
10. Any instrument which has failed or is suspected (or known) to be out of calibration shall be withdrawn from use and labelled conspicuously to prevent accidental use.
11. Adjustable devices shall be sealed to prevent tampering.

1.4 Interpretation of requirements

The general procedures imposed by *BS 5750* for meeting quality assurance targets are fairly easy to understand. In order to comply with the standard, it is necessary to institute and maintain a quality control and measurement system which ensures that the quality of manufactured goods or services does not deviate outside stated error bounds. The *BS 5750* standard itself lays down no demands as to what those error bounds should be. In the case of some products and services, statutory standards apply which are legally enforceable. For example, weighing scales used for commercial purposes have to comply with published standards of accuracy. In other cases, consensus standards apply, often drawn up by trade associations. These represent a general agreement about what level of quality is acceptable in any individual case. Separate national standards (British Standards in the United Kingdom) are often drawn up which express in written form what has been generally agreed in this way.

In a large number of cases, however, the quantification of quality is left entirely to the discretion of the manufacturer or service-provider concerned. It is perfectly permissible as far as the *BS 5750* standard is concerned for a brick manufacturer to specify the error bounds for the dimensions of his bricks as ±10 per cent. Such an error level would be unacceptable to customers of course, and particularly to the bricklayer asked to use such bricks! The choice of appropriate error bounds is thus a marketing decision which balances customer demands for high product specifications against the costs inherent in improving product quality.

In such cases, *BS 5750* requires the manufacturer or service-provider to publish the error bounds he has chosen. He must then institute a system which measures the product at suitable intervals of time and ensures that it does not go outside the stated error bounds. In making such measurements, all instruments used must be calibrated at appropriate intervals to ensure the accuracy of the measurements made, in accordance with procedures laid down in the standard. All these procedures for product measurement and instrument calibration must be fully documented, and this documentation must be made available to customers if required.

The major area of difficulty in conforming to these standards is in proper interpretation of the requirements for providing and maintaining measurement and calibration equipment. Instrument calibration ensures that the measuring accuracy of each instrument involved in the measurement process is known over its whole measurement range, when it is used under specified environmental conditions. This knowledge is gained by comparing the output of the instrument under test against the output of an instrument of known accuracy when the same input is applied to both instruments. What this procedure fails to establish is the accuracy of the instrument when it is used in environmental conditions that differ from those in which it is calibrated, because the characteristics of any instrument vary with ambient conditions, as described in Chapter 5.

As instruments are used in practice to make measurements which are used for quality control functions, it is clearly necessary to establish the variation in instrument characteristics that occurs when the instruments are used in environmental conditions which differ from those applying during calibration. In other words, we need to quantify how modifying inputs (environmental condition changes) affect the performance of an instrument. Knowledge of the modifying inputs that exist, and the way in which they affect particular instruments, also allows measurement procedures to be established which minimize the effect of modifying inputs and thereby improve the quality of measurements. This subject is covered in detail in Chapter 5.

The necessary procedure is therefore to use the calibration of the instrument under standard conditions as a baseline, and make appropriate modifications to account for the variation in characteristics due to differing environmental conditions existing in its normal place of usage. Having done this, it is then possible to quantify the quality of measurements made by calculating the bounds of measurement error of each instrument as it is used in its normal operating environment.

Unfortunately, the quality of measurements does not remain at a constant level forever because the characteristics of any instrument changes over a period of time. The magnitude and rate of this change are influenced by many factors which are difficult or impossible to predict theoretically. It is therefore necessary to apply practical experimentation to determine the rate of such changes in instrument characteristics over a period of time.

Once the maximum permissible measurement error has been defined, knowledge of the rate at which the characteristics of an instrument change allows a time interval to be calculated which represents the moment in time when an instrument will have reached the bounds of its acceptable performance level. The instrument must be re-calibrated either at this time or earlier. This measurement error level which an instrument reaches just before re-calibration is the error bound which must be quoted in the documented specifications for the instrument.

Another requirement is to calculate the cumulative error in a measurement system when the measurement obtained is derived from the output of more than one instrument. Appropriate procedures for this are considered in Chapter 7.

Finally, the whole calibration procedures must be fully documented, in the manner indicated in Chapter 3.

These necessary steps can be summarized as follows:

1. to calibrate all measuring instruments under specified environmental conditions so that their measuring accuracy is known over the whole measurement range;
2. to establish the variation in instrument characteristics under environmental conditions different from the calibration conditions, i.e. to establish how modifying inputs affect the performance of the instrument;
3. to establish measurement procedures which minimize the effect of modifying inputs on the measuring instruments;
4. to calculate the bounds of measurement error of an instrument under normal operating conditions;
5. by practical experimentation to determine the rate of change of instrument characteristics over a period of time;
6. hence to determine the frequency at which instruments should be re-calibrated and the maximum possible measurement error when the instrument has drifted farthest from its specification immediately before calibration;
7. to combine all instrument measurement error levels into a figure which expresses the cumulative error level of the whole measurement process (where a measurement is composed of the outputs of more than one instrument);
8. to document all measurement and calibration procedures.

1.5 Standard measurement units

Before going on to discuss calibration procedures in detail, it is useful to mention the importance of applying standard measuring units. A little of the historical development of standard units is also of interest. In the early stages of human civilization, simple measurement systems were developed to regulate barter trade.

These early units of measurement consisted of convenient parts of the human torso such as the foot and hand. This system allowed an approximate level of equivalence to be established about the relative value of quantities of different commodities. The breadth and width of constructional timber could be measured in units of hands, and its length in units of feet; and so its value for barter trade could be established, for instance relative to cloth measured in units of feet squared. However, this was clearly unfair when someone with large hands and feet exchanged commodities with someone with small hands and feet.

There was an urgent need to establish units of measurement which were based on non-varying quantities, and the first of these was a unit of length (the metre) defined as 10^{-7} times the polar quadrant of the earth. A platinum bar made to this length was established as a standard of length in the early part of the nineteenth century. This was superseded by a superior quality standard bar in 1889, manufactured from a platinum–iridium alloy. Since that time, technological research has enabled further improvements to be made in the accuracy of the standard used for defining length. Firstly, in 1960, a standard metre was re-defined in terms of $1.650\ 763\ 73 \times 10^{6}$ wavelengths of the radiation from Krypton 86 in vacuum. More recently, in 1983, the metre was re-defined yet again as the length of path travelled by light in an interval of $1/299\ 792\ 458$ seconds.

In a similar fashion, standard units for the measurement of other physical quantities have been defined, and the accuracy with which they can be measured has improved progressively over the years. The latest standards for defining the units used for measuring the fundamental physical quantities of length, mass, time, temperature, electric current, luminous intensity and matter are given in Table 1.1. Such standard definitions of these quantities act as primary standards and form the foundation on which all calibration procedures depend. They represent the highest standard of accuracy which is achievable in the measurement of each quantity.

The early establishment of standards for the measurement of physical quantities proceeded in several countries at broadly parallel times and, in consequence, several sets of units emerged for measuring the same physical variable. For instance, length can be measured in yards or metres or several other units. Apart from the major units of length, sub-divisions of standard units exist such as feet, inches, centimetres and millimetres, with a fixed relationship between each fundamental unit and its sub-divisions.

Yards, feet and inches belong to the Imperial System of units, which is characterized by varying and cumbersome multiplication factors relating fundamental units to sub-divisions such as 1760 (miles to yards), 3 (yards to feet) and 12 (feet to inches). This Imperial System is the one used predominantly in Britain until a few years ago.

The metric system is an alternative set of units which includes for instance

Table 1.1 Definitions of standard units

Physical quantity	Standard unit	Definition
LENGTH	metre	The length of path travelled by light in an interval of 1/299 792 458 seconds
MASS	kilogram	The mass of a platinum–iridium cylinder kept in the International Bureau of Weights and Measures, Sèvres, Paris
TIME	second	$9.192\,631\,770 \times 10^9$ cycles of radiation from vapourized caesium 133 (an accuracy of 1 in 10^{12} or 1 second in 36 000 years)
TEMPERATURE	degrees Kelvin	The temperature difference between absolute zero and the triple point of water is defined as $273.16°$ Kelvin
CURRENT	Ampere	One ampere is the current flowing through two infinitely long parallel conductors of negligible cross-section placed 1 metre apart in a vacuum and producing a force of 2×10^{-7} Newtons per metre length of conductor
LUMINOUS INTENSITY	candela	One candela is the luminous intensity in a given direction from a source emitting monochromatic radiation at a frequency of 540 terahertz ($Hz \times 10^{12}$) and with a radiant density in that direction of 1.4641 mW/steradian. (1 steradian is the solid angle which, having its vertex at the centre of a sphere, cuts off an area of the sphere surface equal to that of a square with sides of length equal to the sphere radius)
MATTER	mole	The number of atoms in a 0.012 kg mass of carbon 12

the unit of the metre and its centimetre and millimetre sub-divisions for measuring length. All multiples and sub-divisions of basic metric units are related to the base by factors of ten and such units are therefore much easier to use than Imperial units. However, in the case of derived units such as velocity, the number of alternative ways in which these can be expressed in the metric system can lead to confusion.

As a result of this, an internationally agreed set of standard units (SI units or Systèmes Internationales d'Unités) has been defined, and strong efforts are being made to encourage the adoption of this system throughout the world. In support of this effort, the SI system of units will be used exclusively in this book.

The full range of fundamental SI measuring units and the further set of units derived from them are given in Appendix 1. Two supplementary units for measuring angles are also included within this first appendix. Following this, conversion tables relating common Imperial and metric units to their equivalent SI units are to be found in Appendix 2.

2

Quality control and quality assurance

High standards of product quality and control are essential factors in determining a company's ability to compete in international markets. In order to achieve these goals, administrative and technical system networks must be developed which establish a product quality control system and ensure that the required standards are met and maintained. It is important that these networks identify all activities in a manufacturing system which impact on product quality, and combine to produce a cohesive and effective product quality control system. After such a system has been put into operation, it is also extremely important to monitor its continued effectiveness by generating and evaluating quality-related data during the manufacturing process.

If a quality control system is to be established, then it is essential that commitment to the quality policy is given by all the departments in a company. Full implementation of the quality control policy by all functions in a company is essential: merely having a quality control policy in the form of a document on a shelf is no use whatsoever if it is not put into practice. Without such a total commitment, the potential gains from a quality control system will be lost. Indeed, the company will suffer a double loss, for not only will it fail to grasp the financial benefits accruing from quality control, but it will also have wasted further money in terms of the cost of developing the unused quality control system.

Before proceeding further, some definition of the terminology employed is useful. The three activities necessary to maintain customer satisfaction and a high profile in the market place can be described as quality control, quality management and quality assurance. While these are described in this chapter with reference to manufactured products, they apply equally well to the production of computer software systems and to the provision of services. *Quality* defines the fitness for purpose and conformity with specification of a product. *Quality control* comprises the operational techniques and programmes which are implemented to achieve and sustain the required product quality. *Quality management* describes the managerial aspects of implementing the quality control system and the procedures involved in monitoring and maintaining the effectiveness of such systems. *Quality assurance* is a global term which embraces all functions and activities involved in quality control, both technical and managerial. The purpose of quality assurance is to act as a communication channel, whereby the customer is made aware of the quality control system associated with a particular product and is assured of its effectiveness in maintaining product quality. Documentation is therefore an essential part of quality assurance, wherein all procedures for quality control and management are expressed. Such documentation must be freely available to customers on request, and indeed this availability is specified as one of the conditions to be met in satisfying *BS 5750*.

A useful overall guide to this subject can be found in reference 1.

2.1 Quality policy and objectives

The first step in developing a quality control system must be to establish the general customer needs and expectations. Caution must be exercised, however, in responding to such customer demands. Due regard must be taken of the cost involved in meeting the quality levels demanded by customers. The minimum acceptable quality level can normally be set as that achieved by competing manufacturers in the market place or as defined by published quality standards established by consensus agreement for particular types of product. Improvements in quality beyond this level obviously have strong product-marketing advantages, but only if the cost of better quality levels is not reflected in a significantly higher product price. It is therefore extremely important to have knowledge of the quality level–cost relationship and to define target quality standards appropriately. Figure 2.1 shows the typical relationship between quality level and cost. Assuming that quality control procedures are better than those of competitors in the market place, the manufacturer has the freedom to pitch the target quality level and cost anywhere within the shaded area shown in Figure 2.1 between limits of equal quality level to competing products and equal price, i.e. he can sell a product of better quality at the same price or a product of the same quality at a lower price.

As well as ensuring the quality of a product immediately after delivery to a customer, the objectives of quality control procedures must also include consideration of the reliability of the product over its working life. Good design procedures and adequate quality control procedures during manufacture are essential elements in achieving high reliability in a product. It is also important to make efforts at the design stage of a product to ensure that faults which do

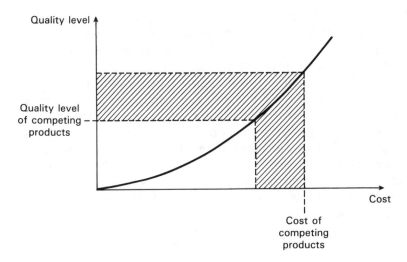

Figure 2.1 Quality level–cost relationship

eventually occur can be quickly and easily rectified. The time taken to repair a product is just as important as its fundamental reliability in many circumstances. The subject of reliability is considered further in Chapter 8.

Legislation about mandatory quality control standards and product liability is obviously an additional factor which must be taken into account in shaping the quality control system developed. Such statutory requirements vary from country to country and due regard for these differences must be taken by any company involved in, or likely to become involved in, export markets. While the fines imposed as a penalty for failing to comply with such legislation might be seen as only a minor irritation which does not have any significant effect on the financial wellbeing of a company, a very much more serious penalty is involved in respect of the inevitable loss of reputation that a company would suffer when prosecuted.

Once the quality level targets have been defined, a strategic plan can be formulated for attaining this required quality level. This will become established as a company policy document and must state clearly what the purpose of the quality control system is, i.e. to ensure the consistency of a product in meeting public interests, laws and regulations and to provide total customer satisfaction in so far as cost factors allow. The strategic plan should include mechanisms for assuring product reliability and meeting promised delivery dates as well as controlling initial product quality, as these are also important parameters in maintaining goodwill and customer satisfaction. A reputation for prompt delivery of high-quality and reliable products will not only keep the current customer base of a company happy but will also gain new customers rapidly.

One very important point about the establishment of quality control systems and the development of a strategic plan to implement them is that these must never be allowed to become stagnant. Technological change brings about continual improvements in manufacturing systems and it is essential to take full account of these. As manufacturing efficiency increases, so the quality level attainable at a given cost also increases. As technological developments make improved quality levels economically viable, a company's quality control system must evolve to improve product quality to this new level. Failure to apply the latest technology and take advantage of the improved product-quality levels which such developments generate must rapidly decrease a company's market competitiveness, as it can be assumed that most competitors of a company in the market place will be implementing such manufacturing system and quality control improvements.

2.2 The quality plan

Quality control must be implemented by a carefully designed and properly documented strategic plan. This will ensure that all quality-related work activities will be organized in a logical fashion and will contribute effectively towards meeting the company's stated product-quality objectives. By using such a strategic plan, the required quality levels will be established and maintained at minimum cost.

The emphasis in the quality plan should be on designing procedures which prevent quality problems occurring rather than on correcting defects after they have occurred. To meet this requirement, it is necessary to design quality and reliability into a product rather than trying to 'bolt it on' afterwards. Prevention is always better than cure!

As far as the technical aspects of quality control are concerned, this means that the requirements to be satisfied must be considered at the design stage of a product. This would not be possible of course if a quality control system were being implemented for an existing product. However, it is sometimes feasible, even in such cases, to make design changes to a product which make achievement and maintenance of a particular quality level easier.

The design specifications of a product are heavily constrained by customer requirements. No degree of conformance to a particular quality level is going to satisfy a customer if the basic design of the product means that it is unable to fulfil everything that the customer wants. If a bus operator orders a fleet of buses seating eighty passengers, he is unlikely to be satisfied if vehicles with a capacity of seventy are supplied, whatever level of reliability and fuel efficiency is claimed for them.

While accepting that product specifications must therefore meet customer demands, various design modifications are often possible which enable quality levels to be met more easily during the manufacturing processes without affecting the product's ability to perform its required functions. It is these possibilities which must be investigated at the design stage. Subsequently, therefore, when the manufacturing process is being planned, it can be assumed that the product involved has been designed in such a way as to allow the required quality levels to be achieved most easily during its manufacture, subject to satisfying the constraint that the product must be able to carry out all the functions required by the customer.

The essential activities in establishing a quality plan are shown schematically in Figure 2.2. The first stage is to identify what the quality objectives are. Specifying the target quality level for a product requires customer

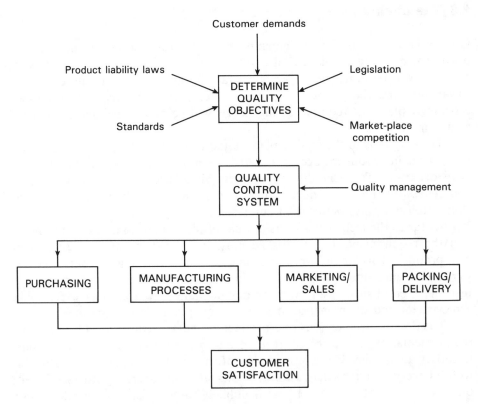

Figure 2.2 Quality plan activities

demands to be carefully balanced against quality improvement costs, while having due regard to what a company's competitors in the market place are doing. A fuller assessment of the costs involved in quality control are given in Section 2.3. Account must also be taken of product liability laws, which require that 'all reasonable steps be taken to ensure the quality of a product'.

Having defined the quality level required, the next step involves careful analysis of the manufacturing process to determine how this target quality level can be achieved and maintained at minimum cost. To do this properly, the whole manufacturing process must be broken down into separate elements. The potential contribution of each element towards achieving the required product quality level must be considered and a quality control plan for each element established accordingly.

In this context, the 'manufacturing process' must be regarded as encompassing the whole activities of a manufacturing company, including such functions as purchasing, marketing, personnel, finance, legal and secretarial, as well as the technical aspects of production. Each of these elements in a company's organizational structure must have clearly defined quality objectives and strategies which contribute towards a cohesive whole policy of quality control and customer satisfaction. Only when such a corporate strategy of quality control is instigated will the full potential benefits of quality management be realized. This is particularly important at any contact points between the company and customers. The impression given by individuals can do much good, or much harm, to the company image. If secretarial staff do not deal promptly, efficiently and courteously with customers, for example, especially with respect to telephone calls, then orders can easily be lost and the benefits of good quality control in technical aspects diminished. The technical aspects of quality control cannot be considered in isolation, therefore, but must be part of a cohesive overall management plan. In other words, quality management should not be considered as being fundamentally different from any other area of management, except in so far as the details of its practical implementation differ. The standard management procedures of planning, organizing, directing and controlling should be applied equally in quality management as in any other management function. This allows quality to be managed in an effective way, whereby planning provides a basis for organizing which in turn enables the quality control system established to be directed and controlled in an efficient manner.

Assurance about the quality of components and raw materials bought in from suppliers lower down the production chain is an essential extension of any quality control system operating in a factory. It is no good instituting careful quality control systems to cover the manufacturing operations taking place within the factory if the raw components coming in from another supplier at the start of the process might be defective.

This requirement is most easily met if the raw component supplier also operates a certified (e.g. to *BS 5750*) quality assurance system. Indeed, in industries such as aerospace, it is mandatory that all components be traceable to a chain of suppliers further down the production line who are all certified as operating an approved quality assurance system. In the absence of such quality certification by suppliers, documented sampling procedures must be established to check the quality of all incoming materials.

The identification of all the separate activities in a company which can affect the attainment of a quality goal is an essential part of the quality control philosophy. However, when the quality control procedures for these separate elements are designed, care must be taken to take full account of their interaction with the rest of the manufacturing system rather than considering them in isolation. Efforts to reduce the cost of quality control in particular parts of a manufacturing system must not simply transfer the cost elsewhere. Failure to take account of this coupling between elements of a quality control system could, for example, result in efforts to reduce inspection costs causing a much greater increase in failure costs. Savings must therefore be real, not imaginary!

In designing a quality control system with a dual purpose of improving product quality levels and minimizing production costs, the size of elemental costs is very relevant. It is much more advantageous to pursue a small reduction in a large cost than a large reduction in a small cost. With this reasoning, it is clear that some costs might be regarded as insignificant and insufficient to justify spending money in trying to reduce them. What that threshold cost level ought to be is of course a subjective decision. However, a clear guideline emerges that efforts to make large cost savings should be instituted before the cost-effectiveness of smaller potential savings is investigated.

Finally, mention must be made of the need for the quality plan to include specification of the planned program of measurements designed to ensure that the required quality is being met and maintained. Such control of product quality at each stage of the manufacturing process is an essential part of the strategic quality control plan. The details of all inspection procedures specified, their prescribed frequency and the measurement techniques required must also be fully documented.

2.3 Quality control system costing

An essential factor in designing a quality control system is the ability to make an accurate assessment of the costs involved in developing and operating the system. While good quality is necessary to maintain customer satisfaction and goodwill, product price is also very important. It is no use making a very high-quality product if its price is too high. The target quality level, therefore, has to be carefully chosen having regard to the cost in achieving that level. For this to be possible, the costs of designing and implementing all aspects of the quality control system must be accurately known. These are summarized in Figure 2.3.

Before going on to discuss quality control system costs in detail, two general points need to be established. Firstly, the accounting system used to assess quality control costs must be appropriate. Some accounting practices can

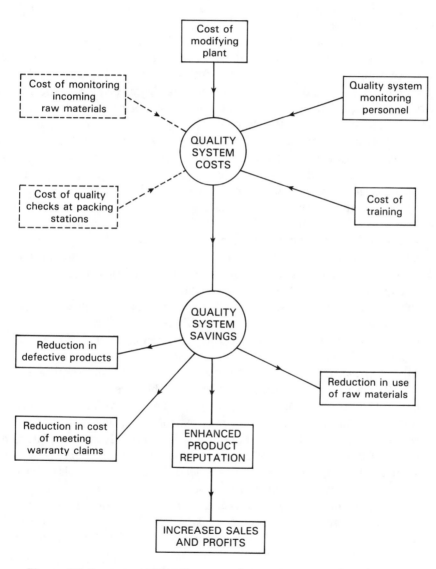

Figure 2.3 Summary of quality control system costs and savings

grossly distort the sizes of both absolute and relative quality control costs by including full overheads in direct labour charges. To swamp the quality control procedure costs in this way with unjustified overheads is quite incorrect. Such ill-thought-out accounting practices can often make a quality control system appear uneconomic, resulting in its non-implementation. The consequence of this is a net loss to the company of the financial benefits that the quality control system would bring if costed sensibly. If the value of a quality control system is to be appraised properly, therefore, only the marginal costs of such labour should be considered.

Secondly, it must be recognized that quality control system costs and the savings generated are not necessarily static. Changes can occur in both over a period of time and thus much of the following discussion about estimating quality control costs and savings should be regarded as being procedures which are repeated at regular intervals of time rather than being 'once only' exercises. Such periodic recalculation should be carried out in recognition of the fact that quality control system costing has relevance well beyond the initial cost estimates used to justify the implementation of the system. Estimates of the costs and savings associated with a quality control system must continue to be generated so that the importance of quality control activities can be conveyed to company managements in meaningful, cost-related terms. Such regularly updated cost information is also a necessary foundation on which continuing improvements in quality control systems can be built.

The most immediately apparent cost in a quality control system is the technical cost of re-designing and modifying manufacturing plant to optimize product quality. To quantify this cost requires a long and in-depth study by experienced engineering personnel of the technical points involved. Other direct quality control costs are the costs of employing additional persons to fill established posts in quality system monitoring and management and the costs of training relevant personnel.

Further quantification of quality control costs becomes more difficult. Many cost factors are in a 'grey area' because they are generated for reasons other than the pursuance of good quality. Purchasing personnel already have a natural function in a company but may expend additional time and effort to ensure that components bought in are of good quality. How should such effort be costed? Another example is at the opposite end of the production cycle where finished components are being packed. Personnel employed to pack components will have been given a mandate to check them for defects as they are packed. This obviously takes a finite amount of extra time. How do you cost that time? Apportionment of such costs is very difficult in the usual situation where a detailed breakdown of how employees spend their time is unavailable.

The expense of making modifications to the design of a manufacturing process to make the achievement of defined quality levels easier can often be a further 'grey-area' cost. If a new plant is being designed, rather than an existing one modified, then the incremental cost of such design modifications cannot be easily quantified. Accounting procedures will obviously vary with regard to if and how these cost components are taken into account. Fortunately, this is not too serious a difficulty in costing quality control properly, as these sorts of costs are relatively insignificant.

An important criterion to meet in following procedures for costing quality control systems is that cost estimates should be accurate, which in turn requires that the underlying data on which the cost estimates are based is accurate. To assess the accuracy of such data, some knowledge about how it was obtained is necessary. Apportionment of an employee's time between quality-related and other functions is particularly prone to inaccuracy, especially when the employee concerned is made responsible for defining this apportionment via time sheets, as the data is then reliant upon his personal, subjective judgement about what work is quality-related and what isn't.

Quality control costs can therefore be divided into those which can be accurately estimated and those which cannot. Both categories of cost should be reported in any viability study or performance record of quality control systems, because otherwise the study could be misleading in underestimating costs without any indication of the existence of errors. Mention of unquantified costs is also useful because it keeps them in view and encourages future attempts to measure them. However, with the caveat that bad estimates are counterproductive, attempts should always be made to estimate all costs if at all possible, as failure to present a full picture is an irritation to those mandated to take a decision about the viability of a proposed quality control system. The credibility of any quality control system proposed is diminished in the absence of a full costing.

Credibility of cost estimates is extremely important because, without credibility, the whole argument for implementing a quality control system is seriously weakened. Extreme care in cost estimates is therefore vital because the exposure of a single weakness in any elemental part of the estimate can undermine confidence in the whole. For this reason costs should be produced by, or at least endorsed by, a company's accounts department, as this puts the necessary 'stamp of approval' on the estimates. In view of their crucial importance, independent corroboration of cost estimates, perhaps by consulting engineers, should also be sought.

The other side of the equation, i.e. an assessment of the financial benefits accruing from implementing quality control procedures, is even more difficult to quantify. The three major elements contributing to financial savings are the production costs of sub-standard products, the value of raw materials wasted in such sub-standard products and the increased sales generated by a reputation for good quality. It would appear that the first two of these elements can be calculated fairly easily. However, if sub-standard components have got as far as the customer, then accounting and purchasing functions in a company become involved in processing the rejects. As this cost is less easy to quantify, the savings arising from reducing reject levels are equally ill-defined.

The third cost-saving element is also difficult to quantify. A simplistic way of measuring increased sales would be to monitor the sales figures over a period of time before and after the introduction of a quality control system. However, this fails to take account of other significant factors, such as whether the total market for a product expands or contracts during the periods in which sales are considered. Quality control systems can only serve to maintain or increase the share which a particular manufacturer has in a market. However, the total size of the market which competing manufacturers are sharing is often outside their control. If the total market contracts, some decrease in sales may occur even with a quality control system in operation. The important point, though, is that the decrease would have been even greater without the quality control system.

When a decision is made about whether to implement a quality control system at all, or what level of quality to aim for, the costs and savings estimates for the system have to be carefully balanced. Comment has already been made about the required credibility for cost estimates, and similar arguments apply to the savings estimates. As for cost estimates, the accuracy of the underlying data on which savings estimates are based is of vital importance. Sloppy procedures in such areas as the generation and processing of rejection notes at the inspection stage, which may cause loss of data, can seriously undermine the accuracy and credibility of savings estimates.

Any product warranties in existence impose yet a further complication in the quality control system cost–savings equation because of the large time-gap between implementing the quality control system and the end of the warranty period when the reduction in warranty claims can be accurately calculated. A similar large time-delay exists between quality control implementation and its effect on a company's reputation and hence its translation into increased sales.

The conclusion to be drawn from the foregoing discussion is that a very accurate assessment of the viability of quality control systems in purely financial terms is at the very least difficult, and most probably impossible. However, costing of a quality control system is still necesary in order that the incremental cost of quality improvement can be assessed, as this is one of the factors used to determine the target quality level that the quality control system is designed to meet. Quality control system costing is also useful in allowing quality control to be applied to the quality control itself, i.e. it ensures that the quality control system developed is commissioned and operated at the minimum cost. It is often useful in this context to divide costs into the three areas associated with design and modifications aimed at fault prevention, quality inspection and appraisal costs, and the cost of rejected products. This classification is convenient because the three areas correspond in financial terms to investment money (design), running costs (appraisal) and negative profit (rejects).

An alternative approach to justifying the decision to implement a quality control system is to consider what would happen if the system was not implemented. Such non-implementation would quickly lead to a serious loss of customers and probable bankruptcy in a climate where rival manufacturers were continuing to develop and install systems to maintain and improve product quality. Justifying expenditure on a quality control system on these grounds makes a full analysis of the implementation and operational costs unnecessary for justification purposes. Some knowledge of costs is still useful for other purposes, however.

Further information on quality costing can be found in reference 2.

2.4 Quality control system implementation

Quality policy must be more than a statement of intent: it must be manifest in a specific course of action designed to achieve the quality objective defined. The quality control system designed must be implemented and maintained in an efficient and effective manner. Key components in doing this (see Figure 2.4) are hardware design and implementation, training, system management, the collection of data to monitor performance, system updates and documentation.

The detailed design and installation of hardware intended to fulfil some

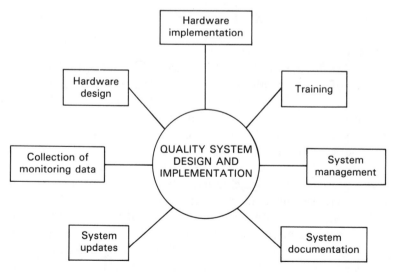

Figure 2.4 Key components in quality control system implementation

quality control function is obviously the responsibility of appropriate engineering personnel. In executing this task, proper management is essential to ensure that the system implemented operates in an efficient manner and performs the functions required of it.

Training relevant personnel is particularly important in establishing and maintaining a quality control system. 'Relevant personnel' should be taken to mean the whole workforce in a company, even if training for some of these only consists of a short one- or two-day course which acquaints them with the quality control procedures operated and their importance.

Management of quality control system implementation and operation has to fulfil several functions. Perhaps the most obvious is its role in ensuring that all hardware and personnel involved in quality control activities operate efficiently and within the cost estimates established at the design stage of the system. Ensuring that all staff involved have been properly trained is a necessary part of this. A second role is to make sure that the quality function interacts properly with all other general company management functions. To accomplish these two roles effectively requires that all the tasks involved are identified and carried out by assigning responsibility, delegating authority and creating accountability for each separate task. As many of the factors contributing to a quality control system such as accounting and purchasing functions lie outside his direct area of control, the quality control manager is necessarily a co-ordinator rather than a manager in many respects. This emphasizes the spirit which must be generated within a company that quality is everyone's concern, not just that of a quality control manager, and by 'everyone' is meant *all* personnel, not just departmental managers.

Organizing the collection of data to monitor system performance is another duty of the quality control system manager. Necessarily included within this is a responsibility for assessing the importance of any deviation of the actual quality measured away from the target quality level. Full knowledge of the importance of such deviations from the customer's viewpoint is necessary in order to arrive at the proper decision about whether to reject the product in question. A proper framework setting out the rules for making this decision must be established, whether the inspection function is performed manually or via an automatic inspection system. This usually involves identifying the various quality characteristics associated with a product, quantifying the various deviations from standard characteristics and giving appropriate weights to indicate the relative importance of each deviation.

Suitable programs of sampling, inspection and testing need to be established for measuring the quality level achieved at all stages in the manufacturing process from the incoming raw materials to the outgoing finished components. The cost of each elemental part of the quality control system must also be monitored continuously, and cost-data collection must be extended beyond the manufacturing cycle to assess the costs involved in products subject to warranty claims. All this information must be presented in a suitable form for analysis and appraisal, such as by bar charts. As far as possible, the information should be combined with production information collected by the production manager to provide an overall management information system. This provides company management with a basis for evaluating the effectiveness of the quality control system and determining the level of compliance with its procedural requirements.

The measurement processes involved in quality-control-inspired data collection are of fundamental importance, and discussion of them is the main purpose of this book. All measurements are derived from instruments of one form or another, and a full knowledge of the possible error level in the data collected is essential. The first requirement for achieving this is that all instruments involved in quality control procedures be calibrated at prescribed intervals of time, as described in Chapter 3. Knowledge of instrument characteristics (Chapter 4), sources of error (Chapter 5), signal processing (Chapter 6), data analysis procedures (Chapter 7) and reliability studies (Chapter 8) are further requirements. The quality control manager must ensure that data is analysed and errors assessed in the correct manner appropriate to the measurements involved.

One factor complicating the design and operation of quality control systems is that the quality levels demanded change continuously under the influence of technological developments and market forces. This requires the quality control system in operation to be updated at various points in time to meet the new demands. Modifications to the system also become necessary if the monitoring exercise of costs and performance shows a deviation from the target cost and performance goals.

2.5 Quality system documentation

Full documentation of all aspects of a quality control system is essential. This should take the form of a quality manual in which every procedure in the quality control system is carefully set out and where the whole philosophy and purpose of the system is described. The quality level target and its justification in terms of balancing the quality costs—savings equation is a very necessary part of this. While there is no particular format defined for this manual, the information must be presented in a clear, systematic and orderly way and it must include all the major elements shown in Figure 2.5. In the case of a small company, all quality system documentation would typically be bound within one manual. For larger companies, however, it is often more appropriate to maintain separate corporate and divisional quality manuals.

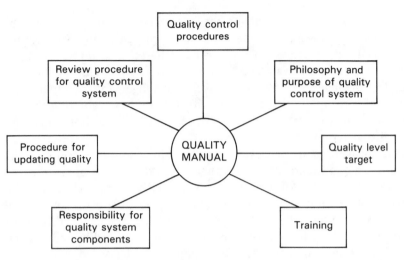

Figure 2.5 Essential components in quality system documentation

Each identified element in the quality control system should be documented such that the details of hardware designs (with full drawings), operational instructions, purpose, running cost and interaction with other elements are all expressed clearly. All documentation should be dated, and this is especially important in the case of engineering drawings of components which are made intermittently in batches. Records must be kept on which version of a drawing was used for any particular batch and what the results of quality control tests on it were. In addition, along with detailed information about the steps taken in the design of a product, all changes made to improve quality must also be documented.

The documentation must include details of any elements of the system which are assigned to sub-contractors. The list of approved contractors must be written down and the conditions regarding the quality of materials supplied, as specified on purchase orders, must also be documented. The exact details of the agreement between purchaser and supplier must be clear so that there is no ambiguity about what quality level is expected.

Review procedures for the quality control system should also be defined in the document and the required frequency for measuring its performance given. Such inspection, testing, auditing, costing and calibration of quality control procedures must be included within the manual as quality records.

Documentation should identify clearly what the organizational structure and mode of operation of the quality control system is, including a description of the associated training procedures. Assignment of responsibility for each elemental part of the system is particularly important and care should be taken to see that activities on the boundaries between functions are adequately managed. The system must be watertight without any loopholes!

Standard procedures for modifying or making additions to the quality manual must also be established and documented within the manual. These must provide for disposing of any parts of the documentation which become redundant or out-dated. To ensure that this procedure functions without mistakes occurring, it is essential that all pages within the quality manual are marked with a date and revision number.

2.6 Quality assurance

The importance of quality assurance in maintaining and expanding the customer base for a product cannot be overemphasized. The whole effort involved in quality control procedures is expended in striving to meet this customer-satisfaction goal. The benefits from the achievement of good quality products, however, are greatly diminished if customers are not made aware of this achievement.

Communication to the customer about the quality control systems associated with a product is therefore an essential part of quality assurance. This is primarily the responsibility of the marketing department in a company but other personnel also have important roles to play in this, such as in the provision of the documentation which demonstrates that the quality control implemented is efficient and maintained satisfactorily. This documentation should show clearly that the quality system conforms to some approved standard (e.g. *BS 5750*) and that associated measurement and calibration procedures follow the guidelines specified. Customers should be able to inspect this documentation at all times and, if necessary, visit the manufacturing plant to see the procedures described in operation.

2.7 Final product inspection

Before the widespread adoption of quality control practices during manufacturing procedures, inspection of products before they were passed to customers was the standard means of ensuring that products delivered were of the required quality. Because measuring the quality of every item produced was prohibitively expensive in most cases, statistical sampling techniques were developed which allowed the overall quality level of a complete production run to be predicted by testing just a few samples from the output. While such statistical methods reduced the cost of product testing, the cost involved was still considerable and unwelcome.

Any system of product inspection merely measures the level of quality: it does not serve any quality control function but exists only to try and prevent defective products being delivered to customers. Because cost factors usually dictate that such product testing must be based on a statistically representative sample of the total number of products in a production run rather than testing every item, there is a finite probability that defective products will not be detected. Final product inspection is therefore expensive to operate and unreliable in its effect.

Quality control procedures have evolved in response to these problems. Their ethos is to manage the manufacturing process in such a manner that the final product always meets the demanded quality levels. This leads to substantial savings in manufacturing costs in comparison with earlier systems of measuring quality by statistical product sampling at the output of the manufacturing process, because the cost of manufacturing defective products which are subsequently discarded at the final inspection point is avoided.

If quality control procedures could be relied upon to operate effectively and efficiently always, then final inspection of products would be totally unnecessary. The quality control procedures would ensure that defects in products could not occur. One day, this situation will be reached, when quality control procedures are implemented by fully automatic means with foolproof fault-detection systems. At the present time, however, quality control systems involve human operation, and the possibility of a deterioration in their effectiveness always exists.

The costs in terms of loss of customer goodwill if quality control procedures break down and defective products are delivered is so enormous that no manufacturer can risk this situation arising. The cost of final product inspection cannot therefore be totally eliminated at the present time, as some system of periodic product testing must continue to be in force as a safeguard against breakdowns in the quality control systems remaining undetected. This product testing would typically need to take the form of periodic quality checks rather than continuous sampling and inspection. Therefore, the cost of product testing with quality control procedures in operation is very much less than that required in the absence of a quality control system.

The optimum method of sampling varies widely with different manufacturing techniques and products. The various sampling procedures available are described elsewhere.[3,4] The sampling procedures described in such texts have usually been developed for full product testing procedures in the era before quality control systems during manufacture became fashionable. Therefore, they need to be applied with intelligence and with appropriate modification to fit them for their new role as a final check on the effectiveness of the quality control system being operated.

Whatever method and frequency of final product testing is instituted, such inspection must be an integral part of the system set up to monitor the continued effectiveness of the manufacturing quality control procedures being operated. This monitoring system must oversee operation of quality control procedures throughout the manufacturing process. It is very important, therefore, that quality be measured at intermediate points in the production process as well as at the final output.

2.8 Quality system audits

If a quality control system is to be operated with the intention of assuring customers of the quality of a product, then those customers must have full confidence in the effectiveness of the quality procedures in operation. Mention has already been made about the importance of allowing customers to have full access to quality system documentation and also of allowing them to make plant visits to see the systems in operation. While such access to documentation and the manufacturing plant is necessary, however, it is not sufficient. In addition, audits of the quality control system being operated must be carried out on a regular basis and recorded in the quality manual to ensure that the system continues to be effective and that no departures have been made from the procedures laid down. Most of these reviews can be carried out by appropriately qualified internal company personnel. However, periodically, an audit must be carried out by an independent third party if the quality system is to have full credibility. Within the United Kingdom, the main provider of such an auditing service is the British Standards Institute: similar bodies exist elsewhere around the world.

Audits by a third party must necessarily be thorough, although if the same audit procedures have been followed by internal reviews (as they should be), then there will be no difficulty in 'passing' the audit and gaining re-certification of the quality control system operated. Such certification of quality control procedures is only required for those parts of a factory for which a company wishes to claim accredited quality assurance. The nature of some products may be such that it is not cost-effective to operate a rigorous quality control system and, in such cases, it is perfectly permissible for a company to register and claim accreditation for quality control procedures that just apply to part of its range of products.

A major item of concern to external auditors will be to satisfy themselves that there is support at the highest level of the company concerned for the quality assurance system operated. Therefore, they will require to have brief discussions with the managing director of the company to confirm this before proceeding with the audit. Thereafter, the starting point of a quality audit is generally to look at the measures taken to ensure the quality of incoming raw materials and part-finished components. Either certification of the quality control procedures operated by the supplier or else an effective sampling method for incoming material will be required. Following this, evidence will be required of the successful operation of the quality control procedures instituted to cover all aspects of the company's activities which have an effect upon product quality. Checks will be made to see that there have been no departures from the documented quality control system (other than agreed ones notified to the accrediting body) and records of the internal audits undertaken to monitor the effectiveness of quality control procedures will be inspected. Where work is being done to a particular published standard (e.g. British Standard Institution Standard), evidence will also be required that the most up-to-date version of that standard is being used. Finally, proof must be shown of the existence of, firstly, an efficient and documented measurement system which monitors quality-related parameters and, secondly, a calibration system traceable to reference standards which check the accuracy of all measuring instruments used.

Quality system auditors will be especially keen to ensure that all personnel whose actions can have an effect on quality are working in the correct manner. They will test this out by selecting people at random to ask them what job they do, what training they have had, where the instructions are that tell them how to do the job and what specific steps they take to maintain product quality. Such questions may be particularly targeted at personnel outside the production departments in areas involving activities such as parts storage, packing and delivery. Such personnel need to know such things as how many components can be stacked on top of each other without causing damage or, in the case of say, delivery of chemicals by tanker, what cleaning procedures are required between loads of different materials.

The outcome of the audit will either be the issue of an accreditation certificate for the quality assurance operated or, alternatively, a statement of what corrective action needs to be taken before the system can be so certified. Once an accreditation certificate is given, arrangements will be made to re-audit the system, perhaps on an annual basis. In between such full audits, unannounced surveillance visits will be made to monitor the continued effectiveness of the quality control procedures agreed.

3

Calibration procedures

Measurements related to product quality are an essential part of quality control systems. Such measurements may be directly related to product quality where they take the form of dimensional measurements, etc., or they may indirectly affect product quality where they take the form of processing temperatures, etc. In either case, accuracy in such measurements is mandatory, and to achieve this accuracy, calibration of the instruments used to obtain the measurements must be carried out at a pre-determined frequency. Such periodic re-calibration is necessary because the characteristics of any measuring instrument change over a period of time and affect the relationship between the input and output. Changes in instrument characteristics are brought about by such factors as mechanical wear, and the effects of dirt, dust, fumes and chemicals in the operating environment. To a great extent, the magnitude of the drift in characteristics depends on the amount of use an instrument receives and hence on the amount of wear and the length of time that it is subjected to the operating environment. However, some drift also occurs, even in storage, as a result of ageing effects in components within the instrument.

Further consideration is given to the reasons for this drift in characteristics in a later chapter on measurement system errors. It is sufficient here to accept that such drift does occur and that the rate at which characteristics change with time varies according to the type of instrument used, the frequency with which it is used and the prevailing environmental conditions. Full knowledge of the mechanisms involved in these two factors is therefore required before the required frequency of instrument re-calibrations can be prescribed.

When calibration is carried out for quality assurance purposes, it is important that all elements in the measurement chain, right up to the final element which produces the quantity used for quality measurement, are calibrated. Thus the signal recorder, as well as all the components within the measuring instrument, must be included in calibration procedures.

3.1 Process instrument calibration

Calibration consists of comparing the output of the process instrument being calibrated against the output of a standard instrument of known accuracy, when the same input (measured quantity) is applied to both instruments. During this calibration process, the instrument is tested over its whole range by repeating the comparison procedure for a range of inputs.

The instrument used as a standard for this procedure must be one which is kept solely for calibration duties. It must never be used for other purposes. Most particularly, it must not be regarded as a spare instrument which can be used for process measurements if the instrument normally used for that purpose breaks down. Proper provision for process instrument failures must be made by keeping a spare set. Standard calibration instruments must be totally separate.

To ensure that these conditions are met, the calibration function must be managed and executed in a professional manner. This will normally mean setting aside a particular place within the instrumentation department of a company where all calibration operations take place and where all instruments used for calibration are kept. As far as possible this should take the form of a separate room, rather than a sectioned-off area in a room that is also used for other purposes. This will enable better environmental control to be applied in the calibration area and will also offer better protection against unauthorized handling or use of the calibration instruments. The level of environmental control required during calibration should be considered carefully with due regard to what level of accuracy is required in the calibration procedure, but should not be overspecified as this will lead to unnecessary expense. Full air conditioning is not normally required for calibration at this level, as it is very expensive, but sensible precautions should be taken to guard the area from extremes of heat or cold, and also good standards of cleanliness should be maintained. Useful guidance on the operation of standards facilities can be found in reference 1.

While it is desirable that all calibration functions are performed in such carefully controlled environment, this is not always practical. Sometimes, it is not convenient or possible to remove instruments from process plant, and in these cases it is standard practice to calibrate them *in-situ*. In these circumstances, appropriate corrections must be made for the deviation in the calibration environmental conditions from those specified. This practice does not obviate the need to protect calibration instruments and maintain them in constant conditions in a calibration laboratory at all times other than when they are involved in such calibration duties on plant.

Apart from the precautions taken to preserve the accuracy of instruments used for calibration, by treating them carefully and reserving them only for calibration duties, they are often chosen to be of a greater inherent accuracy than the process instruments that they are used to calibrate. Where instruments are only used for calibration purposes, greater accuracy can often be achieved by specifying a type of instrument which would be unsuitable for normal process measurements. Ruggedness, for instance is not a requirement, and freedom from this constraint opens up a much wider range of possible instruments. In practice, high-accuracy, null-type instruments are very commonly used for calibration duties, because their requirement for a human operator is not a problem in these circumstances.

As far as management of calibration procedures is concerned, it is important that the performance of all calibration operations is assigned as the clear responsibility of just one person. That person should have total control over the calibration function, and be able to limit access to the calibration laboratory to designated approved personnel only. Only by giving this appointed person total control over the calibration function can the function be expected to operate efficiently and effectively. Lack of such definite management can only lead to unintentional neglect of the calibration system, resulting in the use of equipment in an out-of-date state of calibration and subsequent loss of traceability to reference standards. Professional management is essential so that the customer can be assured that an efficient calibration system is in operation and that the accuracy of measurements is guaranteed.

One of the clauses in *BS 5750* requires that all persons using calibration equipment be adequately trained. The manager in charge of the calibration function is clearly responsible for ensuring that this condition is met. Training must be adequate and targeted at the particular needs of the calibration systems involved. People must understand what they need to know and especially why they must have this information. Successful completion of training courses should be marked by awarding qualification certificates. These attest to the proficiency of personnel involved in calibration duties and are a convenient way of demonstrating that the *BS 5750* training requirement has been satisfied.

The calibration facilities provided within the instrumentation department of a company provide the first link in the calibration chain. Instruments used for calibration at this level are known as *working standards*. A fundamental responsibility in the supervision of this calibration function is to establish the frequency at which the various shop-floor instruments should be calibrated and ensure that calibration is carried out at the appropriate times.

Determination of the frequency at which instruments should be calibrated is dependent upon several factors which require specialist knowledge. If an instrument is required to measure some quantity to an accuracy of ± 2 per cent, then a certain amount of performance degradation can be allowed if its accuracy immediately after re-calibration is ± 1 per cent. What is important is that the pattern of performance degradation be quantified, such that the instrument can be re-calibrated before its accuracy has reduced to the limit defined by the application.

The quantities which cause the deterioration in the performance of instruments over a period of time are mechanical wear, dust, dirt, ambient temperature and frequency of usage. Susceptibility to these factors varies according to the type of instrument involved, and their effect on the accuracy and other characteristics of an instrument can only be quantified by in-depth knowledge of the mechanical construction and other features involved in the instrument. Some form of practical experimentation is normally required to determine the required calibration frequency precisely in the typical operating conditions for the instrument. Further discussion on the means of quantifying the rate of change of instrument characteristics is given in Chapter 5.

A proper course of action must be defined which describes the procedures to be followed when an instrument is found to be out of calibration, i.e. when its output is different to that of the calibration instrument when the same input is applied. The required action depends very much upon the nature of the discrepancy and the type of instrument involved. In many cases, deviations in the form of a simple output bias can be corrected by a small adjustment to the instrument (following which the adjustment screws must be sealed to prevent tampering). In other cases, the output scale of the instrument may have to be redrawn, or scaling factors altered where the instrument output is part of some automatic control or inspection system. In extreme cases, where the calibration procedure shows signs of instrument damage, it may be necessary to send the instrument for repair of even scrap it.

Whatever system and frequency of calibration is established, it is important to review this from time to time to ensure that the system remains effective. It may happen that a cheaper (but equally effective) method of calibration becomes available with the passage of time, and such an alternative system must clearly be adopted in the interests of cost-efficiency. However, the main item under scrutiny in this review is normally whether the calibration interval is still appropriate. Records of the calibration history of the instrument will be the primary basis on which this review is made. It may be that an instrument starts to go out of calibration more quickly after a period of time, either because of ageing factors within the instrument or because of changes in the operating environment. The conditions or mode of usage of the instrument may also be subject to change. As the environmental and usage conditions of an instrument may change beneficially as well as adversely, there is the possibility that the recommended calibration interval may decrease as well as increase.

Maintaining proper records is an important part of fulfilling this calibration function. A separate record, similar to that shown in Table 9.1, should be kept for every instrument in the factory, whether it is in use or kept as a spare. This record should start by giving a description of the instrument and follow this by stating the required calibration frequency. Each occasion on which the instrument is calibrated should be recorded in this record, and every such calibration log should show the status of the instrument in terms of the deviation from its required specification and the action taken to correct it. The calibration record is also very useful in providing feedback which shows whether the calibration frequency has been chosen correctly or not.

3.2 Standards laboratories

We have established so far that process instruments which are used to make quality-related measurements must be calibrated from time to time against a working standard instrument. As this working standard instrument is one which is kept by the instrumentation department of a company for calibration duties, and for no other purpose, then it can be assumed that it will maintain its accuracy over a reasonable period of time because use-related deterioration in accuracy is largely eliminated. However, over the longer term, the characteristics of even such a standard instrument will drift, mainly due to ageing effects in components within it. Over this longer term, therefore, a program must be instituted for calibrating this working standard instrument against one of yet higher accuracy at appropriate intervals of time. The instrument used for calibrating working standard instruments is known as a secondary reference standard. This must obviously be a very well-engineered instrument which gives high accuracy and is stabilized against drift in its performance with time. This implies that it will be an expensive instrument to buy. It also requires that the environmental conditions in which it is used are carefully controlled in respect of ambient temperature, humidity, etc.

When the working standard instrument has been calibrated by an authorized standards laboratory, a calibration certificate will be issued.[2] This will contain at least the following information:

- the identification of the equipment calibrated;
- the calibration results obtained;
- the measurement uncertainty;
- any use limitations on the equipment calibrated;
- the date of calibration;
- the authority under which the certificate is issued.

Establishing a company standards laboratory to provide a calibration facility of the required quality is economically viable only in the case of very large companies where large numbers of instruments need to be calibrated across several factories. In the case of small- to medium-size companies, the cost of buying and maintaining such equipment is not justified. Instead, they would normally use the calibration service provided by various companies which specialize in offering a standards laboratory. What these specialist calibration companies effectively do is to share out the high cost of providing this highly accurate but infrequently used calibration facility over a large number of companies. Such standards laboratories are closely monitored by national standards organizations.[3,4]

3.3 Validation of standards laboratories

In the United Kingdom, the appropriate national standards organization for validating standards laboratories is the National Physical Laboratory (in the United States of America, the equivalent body is the National Bureau of Standards). This has established a National Measurement Accreditation Service (NAMAS) which monitors both instrument calibration and mechanical testing laboratories. The formal structure for accrediting instrument calibration standards laboratories is known as the British Calibration Service (BCS), and that for accrediting testing facilities is known as the National Testing Laboratory Accreditation Scheme (NATLAS).

Although each country has its own structure for the maintenance of standards, each of these different frameworks tends to be equivalent in its effect. To achieve confidence in the goods and services which move across national boundaries, international agreements have established the equivalence of the different accreditation schemes in existence. As a result, NAMAS and the similar schemes operated by France, Germany, Italy, the United States, Australia and New Zealand enjoy mutual recognition.

The British Calibration Service lays down strict conditions which a standards laboratory has to meet before it is approved. These conditions control laboratory management, environment, equipment and documentation. The person appointed as head of the laboratory must be suitably qualified, and independence of operation of the laboratory must be guaranteed. The management structure must be such that any pressure to rush or skip calibration procedures for production reasons can be resisted. As far as the laboratory environment is concerned, proper temperature and humidity control must be provided, and high standards of cleanliness and housekeeping must be maintained. All equipment used for calibration purposes must be maintained to reference standards, and supported by calibration certificates which establish this traceability. Finally, full documentation must be maintained. This should describe all calibration procedures, maintain an index system for re-calibration of equipment, and include a full inventory of apparatus and traceability schedules. Having met these conditions, a standards laboratory becomes an accredited laboratory for providing calibration services and issuing calibration certificates. This accreditation is reviewed at approximately twelve-monthly intervals to ensure that the laboratory is continuing to satisfy the conditions for approval laid down.

3.4 Primary reference standards

Primary reference standards, as listed in Table 1.1, describe the highest level of accuracy that is achievable in the measurement of any particular physical quantity. All items of equipment used in standards laboratories as secondary reference standards have themselves to be calibrated against primary reference standards at appropriate intervals of time. This procedure is acknowledged by the issue of a calibration certificate in the standard way. National standards organizations maintain suitable facilities for this calibration, which in the case of the United Kingdom are at the National Physical Laboratory. The equivalent national standards organization in the United States is the National Bureau of Standards. In certain cases, such primary reference standards can be located outside national standards organizations. For instance, the primary reference standard for dimension measurement is defined by the wavelength of the orange–red line of Krypton light, and it can therefore be realized in any laboratory equipped with an interferometer.

In certain cases (e.g. the measurement of viscosity), such primary reference standards are not available and reference standards for calibration are achieved by collaboration between several national standards organizations who perform measurements on identical samples under controlled conditions.[5,6]

3.5 Traceability

What has emerged from the foregoing discussion is that calibration has a chain-like structure in which every instrument in the chain is calibrated against a more accurate instrument immediately above it in the chain, as shown in Figure 3.1. All the elements in the calibration chain must be known so that the calibration of process instruments at the bottom of the chain is traceable to the fundamental measurement standards.

This knowledge of the full chain of instruments involved in the calibration procedure is known as *traceability*, and is specified as a mandatory requirement in satisfying standards such as *BS 5750*. Documentation must exist which shows that process instruments are calibrated by standard instruments which are

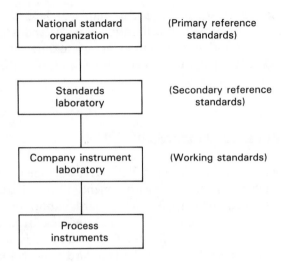

Figure 3.1 Instrument calibration chain

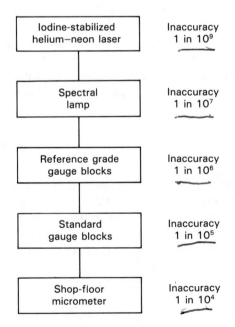

Figure 3.2 Typical calibration chain for micrometers

linked by a chain of increasing accuracy back to national reference standards. There must be clear evidence to show that there is no break in this chain.

To illustrate a typical calibration chain, consider the calibration of micrometers (Figure 3.2). A typical shop-floor micrometer has an uncertainty (inaccuracy) of less than 1 in 10^4. These would normally be calibrated in the instrumentation department laboratory of a company against laboratory standard gauge blocks with a typical uncertainty of less than 1 in 10^5. A specialist calibration service company would provide facilities for calibrating these laboratory standard gauge blocks against reference grade gauge blocks with a typical uncertainty of less than 1 in 10^6. More accurate calibration equipment still is provided by national standards organizations. The National Physical Laboratory maintains two sets of standards for this type of calibration: a working standard and a primary standard. Spectral lamps are used to provide a working reference standard with an uncertainty of less than 1 in 10^7. The primary standard is provided by an iodine-stabilized helium–neon laser which has a specified uncertainty of less than 1 in 10^9. All the links in this calibration chain must be shown in any documentation which describes the use of micrometers in making quality-related measurements.

3.6 Documentation in the workplace

An essential element in the maintenance of measurement systems and the operation of calibration procedures is the provision of full documentation. This must give a full description of the measurement requirements throughout the workplace, the instruments used, and the calibration system and procedures operated. Individual calibration records for each instrument must be included within this. This documentation is a necessary part of the quality manual, although it may physically exist as a separate volume if this is more convenient. An over-riding constraint on the style in which the documentation is presented is that it should be simple and easy to read. This is often greatly facilitated by the copious use of appendices.

The starting point in the documentation must be a statement of the measurement limits that have been defined for each measurement system documented. Such limits are established by balancing the costs of improved accuracy against customer requirements, and also with regard to the overall quality level specified in the quality manual. The technical procedures required for this, which involve assessing the type and magnitude of relevant measurement errors, are described in Chapters 4–7. It is customary to express the final measurement limit calculated as ±2 standard deviations, i.e. within 95 per cent confidence limits (see Chapter 7 for an explanation of these terms).

The instruments specified for each measurement situation must be listed next. This list must be accompanied by full instructions about the proper use of the instruments concerned. These instructions will include details about any environmental control or other special precautions which must be taken to ensure that the instruments provide measurements of sufficient accuracy to meet the measurement limits defined. The proper training courses appropriate to plant personnel who will use the instruments must be specified.

Having disposed of the question about which instruments are used, the documentation must go on to cover the subject of calibration. Full calibration is not applied to every measuring instrument used in a workplace because *BS 5750* acknowledges that formal calibration procedures are not necessary for some equipment where it is uneconomic or technically unnecessary because the accuracy of the measurement involved has an insignificant effect on the overall quality target for a product. However, any equipment which is excluded from calibration procedures in this manner must be specified as such in the documentation. Identification of equipment in this category is a matter of informed judgement.

For instruments which are the subject of formal calibration, the documentation must specify the standard instruments to be used for the purpose and define a formal procedure of calibration. The procedure must include instructions for the storage and handling of standard calibration instruments and specify the required environmental conditions under which calibration is to be performed. Where a calibration procedure for a particular instrument uses published standard practices, it is sufficient to include a reference to that standard procedure in the documentation rather than to reproduce the whole procedure. Whatever calibration system is established, a formal review procedure must be defined in the documentation which ensures its continued effectiveness at regular intervals. The results of each review must also be documented in a formal way.

A standard format for the recording of calibration results should be defined in the documentation. A separate record must be kept for every instrument present in the workplace which includes details of the instrument's description, the required calibration frequency, the date of each calibration and the calibration results on each occasion. Where appropriate, the documentation must also define the manner in which calibration results are to be recorded on the instruments themselves.

The documentation must specify procedures which are to be followed if an instrument is found to be outside the calibration limits. This may involve adjustment, re-drawing its scale or withdrawing an instrument, depending upon the nature of the discrepancy and the type of instrument involved. Instruments withdrawn will either be repaired or scrapped. In the case of withdrawn instruments, a formal procedure for marking them as such must be defined to prevent them being accidentally put back into use.

Two other items must also be covered by the calibration document. The traceability of the calibration system back to national reference standards must be defined and supported by calibration certificates (see Section 3.2). Training procedures must also be documented, specifying the particular training courses to be attended by various personnel and what, if any, refresher courses are required.

All aspects of these documented calibration procedures will be given consideration as part of the periodic audit of the quality control system which calibration procedures are instigated to support. While the basic responsibility for choosing a suitable interval between calibration checks rests with the engineers responsible for the instruments concerned, the quality system auditor will require to see the results of tests which show that the calibration interval has been chosen correctly and that instruments are not going outside allowable measurement uncertainty limits between calibrations. Particularly important in such audits will be the existence of procedures which are instigated in response to instruments found to be out of calibration. Evidence that such procedures are effective in avoiding degradation in the quality assurance function will also be required.

4

Instrument classification, characteristics and choice

One important aspect of quality control and assurance is the correct choice of instruments for monitoring process variables and measuring product dimensions and other quality-related variables. Knowledge of the possible error levels in measurements is essential, and a necessary pre-requisite for this is a proper understanding of the operational characteristics of instruments and an examination of the way in which instrument performance is specified. A convenient way to achieve this knowledge is to classify instruments into different types and then to study the characteristics of each of these various instrument sub-groups.

Instruments consist of one or more separate components which together serve to give an output reading which is some function of a measured physical quantity. The primary component in an instrument is a transducer which translates the measured physical quantity into another form. Further possible components within the instrument are an amplifier, an amplifier–analyser and an output display system. The term 'instrument' is used somewhat loosely throughout this text, as is fairly common practice, to describe any or all of these components.

4.1 Instrument classification

Instruments can be sub-divided into separate classes according to several criteria. These sub-classifications are useful in broadly establishing several attributes of particular instruments such as accuracy, cost, and general applicability to different applications.

44

Active/passive instruments

Instruments are divided into active or passive ones according to whether the instrument output is entirely produced by the quantity being measured or whether the quantity being measured simply modulates the magnitude of some external power source. This might be more easily understood if it were illustrated by an example.

An example of a *passive instrument* is the pressure-measuring device, shown in Figure 4.1 . The pressure of the fluid is translated into movement of a pointer against a scale. The energy expended in moving the pointer is derived entirely from the change in pressure measured; there are no other energy inputs to the system.

An example of an *active instrument* is a petrol-tank-level indicator, as sketched in Figure 4.2 . Here, the change in petrol level moves a potentiometer arm, and the output signal consists of a proportion of the external voltage source applied across the two ends of the potentiometer. The energy in the output signal comes from the external power source; the primary transducer float system is merely modulating the value of the voltage from this external power source.

In active instruments, the external power source is usually in electrical form, but in some cases it can be other forms of energy such as a pneumatic or hydraulic one.

One very important difference between active and passive instruments is the level of measurement resolution which can be obtained. With the simple pressure gauge shown, the amount of movement made by the pointer for a particular pressure change is closely defined by the nature of the instrument. While it is possible to increase measurement resolution by making the pointer

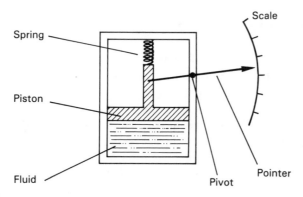

Figure 4.1 Passive pressure gauge

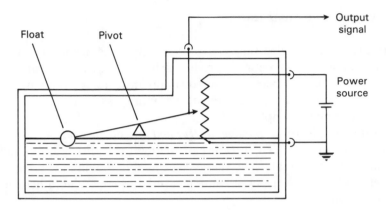

Figure 4.2 Petrol-tank-level indicator

longer, such that the pointer tip moves through a longer arc, the scope for such improvement is clearly bounded by the practical limit on how long the pointer can conveniently be. In an active instrument, however, adjustment of the magnitude of the external energy input allows much greater control over measurement resolution. While the scope for improving measurement resolution is much greater incidently, it is not infinite because of limitations placed on the magnitude of the external energy input, in consideration of heating effects and for safety reasons.

In terms of cost, passive instruments are normally of a simpler construction than are active ones, and are therefore cheaper to manufacture. Choice between active and passive instruments for a particular application thus involves balancing the measurement–resolution requirements carefully against cost.

Null/deflection-type instruments

The last pressure gauge is a good example of a *deflection type* of instrument, where the value of the quantity being measured is displayed in terms of the amount of movement of a pointer.

An alternative type of pressure gauge is the dead-weight gauge shown in Figure 4.3 which is a *null-type* instrument. Here, weights are put on top of the piston until the downward force balances the fluid pressure. Weights are added until the piston reaches a datum level, known as the null point. Pressure measurement is made in terms of the value of the weights needed to reach this null position.

The accuracy of these two instruments depends on different things. For the

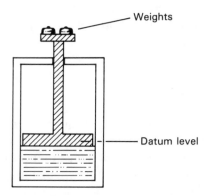

Figure 4.3 Dead-weight pressure gauge

first, it depends on the linearity and calibration of the spring, while for the second, it relies on the calibration of the weights. As calibration of weights is much easier than careful choice and calibration of a linear-characteristic spring, this means that the second type of instrument will normally be the more accurate. This is in accordance with the general rule that null-type instruments are more accurate than deflection types.

In terms of usage, the deflection-type instrument is clearly more convenient. It is far simpler to read off the position of a pointer against a scale than to add and subtract weights until a null point is reached. A deflection-type instrument is therefore the one that would normally be used in the workplace. For calibration duties, however, the null-type instrument is preferable because of its superior accuracy. The extra effort required to use such an instrument is perfectly acceptable in this case because of the infrequent nature of calibration operations.

Monitoring/control instruments

An important distinction between different instruments is made according to whether they are suitable only for monitoring functions or whether their output is in a form that can be directly included as part of an automatic control system. Instruments which only give an audio or visual indication of the magnitude of the physical quantity measured, such as a liquid-in-glass thermometer, are only suitable for monitoring purposes. This class normally includes all null-type instruments and most passive transducers.

For an instrument to be suitable for inclusion in an automatic control system, its output must be in a suitable form for direct input into the controller.

This usually means that an instrument with an electrical output is required, although other forms of output such as optical or pneumatic signals are used in some systems.

Analogue/digital instruments

An *analogue instrument* gives an output which varies continuously as the quantity being measured changes. The output can have an infinite number of values within the range that the instrument is designed to measure. The deflection type of pressure gauge described earlier in this chapter is a good example of an analogue instrument. As the input value changes, the pointer moves with a smooth continuous motion. While the pointer can therefore be in an infinite number of positions within its range of movement, the number of different positions which the eye can discriminate between is strictly limited, this discrimination being dependent upon how large the scale is and how finely it is divided.

A *digital instrument* has an output which varies in discrete steps and so can only have a finite number of values. The rev counter sketched in Figure 4.4 is an example of a digital instrument. In this, a cam is attached to the revolving body whose motion is being measured, and on each revolution the cam opens and closes a switch. The switching operations are counted by an electronic counter. This system can only count whole revolutions and therefore cannot register any motion which is less than a full revolution.

The distinction between analogue and digital instruments has become particularly important with the rapid growth in the application of microcomputers to automatic control systems. Any digital computer system, of which the microcomputer is but one example, performs its computations in digital form. An instrument whose output is in digital form is therefore particularly

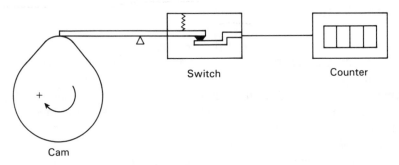

Figure 4.4 Rev counter

advantageous in such applications, as it can be interfaced directly to the control computer. Analogue instruments must be interfaced to the microcomputer by an analogue-to-digital (A/D) converter, which converts the analogue output signal from the instrument into an equivalent digital quantity which can be read into the computer. This conversion has several disadvantages. Firstly, the A/D converter adds a significant cost to the system. Secondly, a finite time is involved in the process of converting an analogue signal to a digital quantity, and this time can be critical in the control of fast processes where the accuracy of control depends on the speed of the controlling computer. Degrading the speed of operation of the control computer by imposing a requirement for A/D conversion thus degrades the accuracy by which the process is controlled.

4.2 Static instrument characteristics

If we have a thermometer in a room and its reading shows a temperature of 20 °C, then it does not really matter whether the true temperature of the room is 19.5 °C or 20.5 °C. Such small variations around 20 °C are too small to affect whether we feel warm enough or not. Our bodies cannot discriminate between such close levels of temperature and therefore a thermometer with an accuracy of ±0.5 °C is perfectly adequate. If we had to measure the temperature of certain chemical processes however, a variation of 0.5 °C might have a significant effect on the rate of reaction or even the products of a process. A measurement accuracy much better than ±0.5 °C is therefore clearly required in this case.

Accuracy of measurement is thus one consideration in the choice of instrument for a particular application. Other quantities such as sensitivity, linearity and the reaction to ambient temperature changes are further considerations. These attributes are collectively known as the 'static characteristics' of instruments, and are given in the data sheet for a particular instrument. It is important to note that the values quoted for instrument characteristics in such a data sheet only apply when the instrument is used under specified standard calibration conditions. Due allowance must be made for variations in the characteristics when the instrument is used in other conditions.

The various static characteristics are defined in the following paragraphs. More formal definitions can be found in references 1, 2 and 3.

Accuracy

Accuracy is the extent to which a reading might be wrong, and is often quoted — as a percentage of the full-scale reading of an instrument.

If, for example, a pressure gauge of range 0–10 bar has a quoted inaccuracy of ±1.0 per cent f.s. (±1 per cent of full-scale reading), then the maximum error to be expected in *any* reading is 0.1 bar. This means that when the instrument is reading 1.0 bar, the possible error is 10 per cent of this value. For this reason, it is an important system design rule that instruments are chosen such that their range is appropriate to the spread of values being measured, in order that the best possible accuracy be maintained in instrument readings. Thus, if we were measuring pressures with expected values between 0 and 1 bar, we would not use an instrument with a range of 0–10 bar.

Tolerance

Tolerance is a term which is closely related to accuracy and defines the maximum error which is to be expected in some value. While it is not, strictly speaking, a static characteristic of measuring instruments, it is mentioned here because the accuracy of some instruments is sometimes quoted as a tolerance figure.

Tolerance, when used correctly, describes the maximum deviation of a manufactured component from some specified value. Crankshafts, for instance, are machined with a diameter tolerance quoted as so many microns, and electric circuit components such as resistors have tolerances of perhaps 5 per cent. One resistor chosen at random from a batch having a nominal value 1000Ω and tolerance 5 per cent might have an actual value anywhere between 950Ω and 1050Ω.

Precision/repeatability/reproducibility

Precision is a term which describes an instrument's degree of freedom from random errors. If a large number of readings are taken of the same quantity by a high-precision instrument, then the spread of readings will be very small.

High precision does not imply anything about measurement accuracy. A high-precision instrument may have a low accuracy. Low-accuracy measurements from a high-precision instrument are normally caused by a bias in the measurements, which is removable by re-calibration.

The terms repeatability and reproducibility mean approximately the same but are applied in different contexts as given below. *Repeatability* describes the closeness of output readings when the same input is applied repetitively over a short period of time, with the same measurement conditions, same instrument and observer, same location and same conditions of use maintained throughout. *Reproducibility* describes the closeness of output readings for the same input when there are changes in the method of measurement, observer, measuring instrument, location, conditions of use and time of measurement. Both terms thus describe the spread of output readings for the same input. This spread is referred to as repeatability if the measurement conditions are constant and as reproducibility if the measurement conditions vary.

The degree of repeatability or reproducibility in measurements from an instrument is an alternative way of expressing its precision. Figure 4.5 illustrates this more clearly. The diagram shows the results of tests on three industrial robots which were programmed to place components at a particular point on a table. The target point was at the centre of the concentric circles shown and

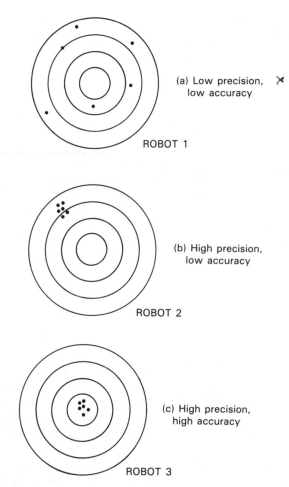

(a) Low precision, low accuracy

ROBOT 1

(b) High precision, low accuracy

ROBOT 2

(c) High precision, high accuracy

ROBOT 3

Figure 4.5 Comparison of accuracy and precision

the black dots represent the points where each robot actually deposited components at each attempt. Both the accuracy and precision of Robot 1 is shown to be low in this trial. Robot 2 consistently puts the component down at approximately the same place but this is the wrong point. Therefore, it has high precision but low accuracy. Finally, Robot 3 has both high precision and high accuracy, because it consistently places the component at the correct target position.

Range or span

The *range* or *span* of an instrument defines the minimum and maximum values of a quantity that the instrument is designed to measure.

Bias

Bias describes a constant error which exists over the full range of measurement of an instrument. This error is normally removable by calibration.

Bathroom scales are a common example of instruments which are prone to bias. It is quite usual to find that there is a reading of perhaps 1 kg with no one standing on the scales. If someone of known weight 70 kg were to get on the scales, the reading would be 71 kg, and if someone of known weight 100 kg were to get on the scales, the reading would be 101 kg. This constant bias of 1 kg can be removed by calibration: in the case of bathroom scales this normally means turning a thumbwheel with the scales unloaded until the reading is zero.

Linearity

It is normally desirable that the output reading of an instrument is linearly proportional to the quantity being measured. The ×s marked on Figure 4.6 show a plot of the typical output readings of an instrument when a sequence of input quantities are applied to it. Normal procedure is to draw a good-fit straight line through the ×s, as shown in this diagram. (While this can often be done with reasonable accuracy by eye, it is always preferable to apply a mathematical least-squares line-fitting technique, as described in Chapter 7.) The *non-linearity* is then defined as the maximum deviation of any of the output readings marked × from this straight line. Non-linearity is usually expressed as a percentage of full-scale reading.

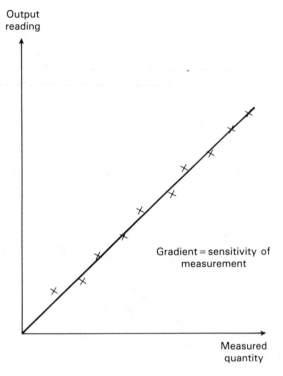

Figure 4.6 Instrument output characteristic

Sensitivity of measurement

The *sensitivity of measurement* is a measure of the change in instrument output which occurs when the quantity being measured changes by a given amount. Sensitivity is thus the ratio

$$\frac{Scale\ deflection}{Value\ of\ measured\ quantity\ causing\ deflection}$$

The sensitivity of measurement is therefore the slope of the straight line drawn on Figure 4.6. If, for example, a pressure of 2 bars produces a deflection of 10° in a pressure transducer, the sensitivity of the instrument is 5°/bar (assuming that the relationship between pressure and the instrument reading is a straight-line one).

Sensitivity to disturbance

All calibrations and specifications of an instrument are only valid under

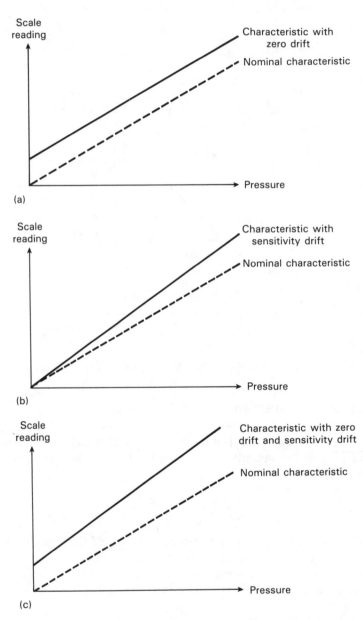

Figure 4.7: (a) zero drift; (b) sensitivity drift; (c) zero drift plus sensitivity drift

controlled conditions of temperature, pressure, etc. These standard ambient conditions are usually defined in the instrument specification. As variations occur in the ambient temperature, etc., certain static instrument characteristics change, and the sensitivity to disturbance is a measure of the magnitude of this change.

Such environmental changes affect instruments in two main ways, known as *zero drift* and *sensitivity drift*.

Zero drift

Zero drift describes the effect where the zero reading of an instrument is modified by a change in ambient conditions.

Typical units by which zero drift is measured are volts/$^\circ$C in the case of a voltmeter affected by ambient temperature changes. The effect of zero drift is to impose a bias in the instrument output readings: this is normally removable by re-calibration in the usual way. A typical change in the output characteristic of a pressure gauge subject to zero drift is shown in Figure 4.7(a).

Sensitivity drift (scale factor drift)

Sensitivity drift defines the amount by which an instrument's sensitivity of measurement varies as ambient conditions change. An alternative name for this phenomenon is scale factor drift.

Many components within an instrument are affected by environmental fluctuations, such as temperature changes: for instance, the modulus of elasticity of a spring is temperature-dependent. Figure 4.7(b) shows what effect sensitivity drift can have on the output characteristic of an instrument. Sensitivity drift is measured in units of the form (angular degree/bar)/$^\circ$C.

If an instrument suffers both zero drift and sensitivity drift sumultaneously, then the typical modification of the output characteristic is shown in Figure 4.7(c).

Hysteresis

Figure 4.8 illustrates the output characteristic of an instrument which exhibits *hysteresis*. If the measured quantity input to the instrument is steadily increased from a negative value, the output reading varies in the manner shown in curve (a). If the input variable is then steadily decreased, the output varies in the

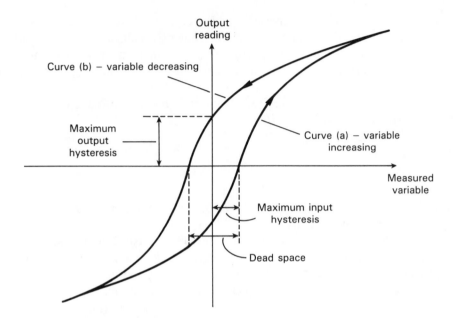

Figure 4.8 Instrument characteristic with hysteresis

manner shown in curve (b). The non-coincidence between these loading and unloading curves is known as *hysteresis*.

Two quantities, *maximum input hysteresis* and *maximum output hysteresis*, are defined as shown in the diagram. These are normally expressed as a percentage of the full-scale input or output reading respectively.

Dead space

Dead space is defined as the range of different input values over which there is no change in output value.

Any instrument which exhibits hysteresis also displays dead space, as marked on Figure 4.8. Some instruments which do not suffer from any significant hysteresis can still exhibit a dead space in their output characteristics, however. Backlash in gears is a typical cause of dead space, and results in the sort of instrument output characteristic shown in Figure 4.9.

Threshold

If the input to an instrument is gradually increased from zero, it will have to reach a certain minimum level before the change in the instrument output

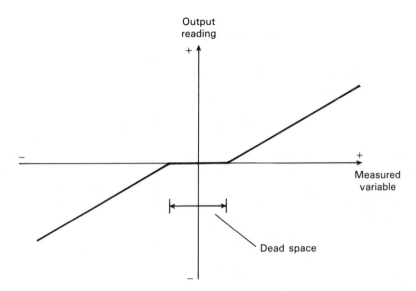

Figure 4.9 Instrument characteristic with dead space

reading is of a sufficiently large magnitude to be detectable. This minimum level of input is known as the *threshold* of the instrument. Manufacturers vary in the way that they specify threshold for instruments. Some quote absolute values, whereas others quote threshold as a percentage of full-scale readings.

As an illustration, a car speedometer typically has a threshold of about 15 km/h. This means that if the vehicle starts from rest and accelerates, no output reading is observed on the speedometer until the speed reaches 15 km/h.

Resolution

When an instrument is showing a particular output reading, there is a lower limit on the magnitude of the change in the input measured quantity which produces an observable change in the instrument output. Like threshold, resolution is sometimes specified as an absolute value and sometimes as a percentage of full-scale deflection.

One of the major factors influencing the resolution of an instrument is how finely its output scale is sub-divided. Using a car speedometer as an example again, this has sub-divisions of typically 20 km/h. This means that when the needle is between the scale markings, we cannot estimate speed more accurately than to the nearest 5 km/h. This figure of 5 km/h thus represents the resolution of the instrument.

4.3 Dynamic instrument characteristics

The static characteristics of measuring instruments are concerned only with the steady-state reading that the instrument settles down to, such as the accuracy of the reading, etc.

The dynamic characteristics of a measuring instrument describe its behaviour between the time that a measured quantity changes value and the time that the instrument output attains a steady value in response. As with static characteristics, any values for dynamic characteristics quoted in instrument data sheets only apply when the instrument is used under specified environmental conditions. Outside these calibration conditions, some variation in the dynamic parameters can be expected.

Various types of dynamic characteristics can be classified, known as zero order, first order and second order characteristics. The practical effects of dynamic characteristics on the output of an instrument can be understood without resorting to formal mathematical analysis. However, this analysis is presented in Appendix 3 for interested readers.

In a *zero order instrument*, the dynamic characteristics are negligible and the instrument output responds and reaches its final reading almost instantaneously following a step change in the measured quantity applied at its input. A potentiometer, which measures motion, is a good example of such an instrument, where the output voltage changes approximately instantaneously as the slider is displaced along the potentiometer track.

In a *first order instrument*, the output quantity q_o in response to a step change in the measured quantity q_i varies with time in the manner shown in Figure 4.10. The time constant τ of the step response is the time taken for the output quantity q_o to reach 63 per cent of its final value.

The liquid-in-glass thermometer is a good example of a first order instrument. It is well known that if a mercury thermometer at room temperature is plunged into boiling water, the mercury does not rise instantaneously to a level indicating $100\,^\circ\text{C}$, but instead approaches a reading of $100\,^\circ\text{C}$ in the manner indicated in Figure 4.10.

A large number of other instruments also belong to this first order class: this is of particular importance in control systems where it is necessary to take account of the time lag that occurs between a measured quantity changing in value and the measuring instrument indicating the change. Fortunately, the time constant of many first order instruments is small relative to the dynamics of the process being measured, and so no serious problems are created.

It is convenient to describe the characteristics of *second order instruments* in terms of three parameters K (static sensitivity), ω (undamped natural frequency) and ε (damping ratio). These are formally defined in Appendix 3.

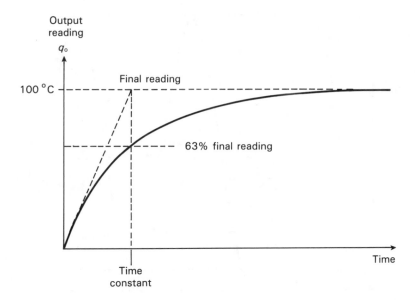

Figure 4.10 First order instrument characteristic

The manner in which the output reading of a second order instrument changes following a change in the measured quantity applied to its input depends on the value of these three parameters. The damping ratio parameter, ε, controls the shape of the output response and responses of a second order instrument for various values of ε are shown in Figure 4.11. For case (a) where $\varepsilon = 0$, there is no damping and the instrument output exhibits constant amplitude oscillations when disturbed by any change in the physical quantity measured. For light damping of $\varepsilon = 0.2$, represented by case (b), the response to a step change in input is still oscillatory but the oscillations gradually die down. Further increase in the value of ε reduces oscillations and overshoot still more, as shown by curves (c) and (d), and finally the response becomes very overdamped, as shown by curve (e) where the output reading creeps up slowly towards the correct reading. Clearly, the extreme response curves (a) and (e) are grossly unsuitable for any measuring instrument. If an instrument were to be only ever subjected to step inputs, then the design strategy would be to aim towards a damping ratio of 0.707, which gives the critically damped response (c). Unfortunately, most of the physical quantities which instruments are required to measure do not change in mathematically convenient steps, but rather in the form of ramps of varying slopes. As the form of the input variable changes, so the best value for ε varies, and choice of ε becomes one of compromise between those values that are best for each type of input variable behaviour anticipated. Commercial second order instruments, of which the

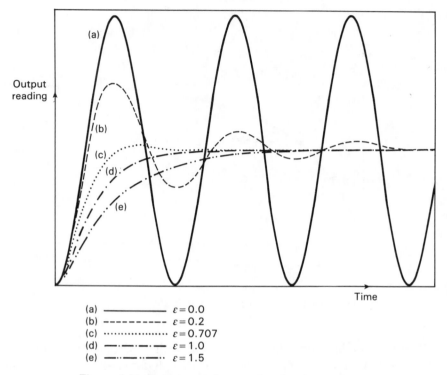

Output reading

Time

(a) ——————— $\varepsilon = 0.0$
(b) ---------- $\varepsilon = 0.2$
(c) ················ $\varepsilon = 0.707$
(d) —·—·—·— $\varepsilon = 1.0$
(e) —··—··—·· $\varepsilon = 1.5$

Figure 4.11 Response of second order instruments

accelerometer is a common example, are generally designed to have a damping ratio (ε) somewhere in the range of 0.6–0.8.

4.4 Cost, durability and maintenance

The static and dynamic characteristics discussed so far are those features which form the technical basis for a comparison between the relative merits of different instruments. In assessing the relative suitability of different instruments for a particular measurement situation, however, considerations of cost, durability and maintenance are also of great importance.

Cost is very strongly correlated with the performance of an instrument, as measured by its characteristics. Increasing the accuracy or resolution of an instrument, for example, can only be done at a penalty of increasing its manufacturing cost. Instrument choice therefore proceeds by specifying the minimum chacteristics required by a measurement situation and then searching manufacturers' catalogues to find an instrument whose characteristics match those required. As far as accuracy is concerned, it is usual to specify maximum measurement uncertainty levels which are 10 per cent of the tolerance levels of the parameter to be measured. To select an instrument whose accuracy and other characteristics are superior to the minimum levels required would only mean paying more than necessary for a level of performance which is greater than that needed.

As well as purchase cost, other important factors in the assessment exercise are instrument durability and maintenance requirements. Assuming that one had £10 000 to spend, it would not be sensible to spend £9000 on a new motor car whose projected life was five years if a car of equivalent specification with a projected life of ten years was available for £10 000. Likewise, durability is an important consideration in the choice of instruments. The projected life of instruments often depends on the conditions in which the instrument will have to operate. Maintenance requirements must also be taken into account, as they also have cost implications.

As a general rule, a good assessment criterion is obtained if the total purchase cost and estimated maintenance costs of an instrument over its life are divided by the period of its expected life. The figure obtained is thus a cost per year. However, this rule becomes modified where instruments are being installed on a process whose life is expected to be limited, perhaps in the manufacture of a particular model of car. Then, the total costs can only be divided by the period of time an instrument is expected to be used, unless an alternative use for the instrument is envisaged at the end of this period.

To summarize, therefore, instrument choice is a compromise between performance characteristics, ruggedness and durability, maintenance requirements and purchase cost. To carry out such an evaluation properly, the instrument engineer must have a wide knowledge of the range of instruments available for measuring particular physical quantities, and he/she must also have a deep understanding of how instrument characteristics are affected by particular measurement situations and operating conditions.

5

Sources of measurement error

The importance of measurements in quality control has been emphasized already in Chapter 2. All manufacturing process parameters and characteristics of a final product which are the subject of quality control must be measured to standards of accuracy which are clearly known. It is also normally required that those standards of accuracy be as high as possible.

The measurement system designer is therefore charged with the task of reducing errors in instrument output readings to the minimum possible level and then quantifying the maximum remaining error which may exist in any output reading. In order to achieve this, a detailed analysis of the sources of error which exist in a measurement system must be carried out.

This chapter is concerned with identifying the various measurement system errors which exist, and suggesting mechanisms for reducing their magnitude and effect. Such errors in measurement data can be divided into random errors and systematic errors. The way in which these separate error components are assessed in order to calculate the overall measurement system error level is considered later in Chapter 7.

5.1 Random and systematic errors

Random errors describe small differences in the output readings of an instrument when the same quantity is measured a number of times. The magnitude and sign of the error is random, so that for a large number of samples, the number of positive errors will be balanced by the number of negative errors and the mean error will be zero. In most systems, random errors have a Gaussian distribution about this zero mean (see Chapter 7). The magnitude of random errors in measurements is quantified by the terms 'repeatability' and 'reproducibility' (formally defined in Chapter 4).

A typical situation in which random errors occur is when measurements are made by human observation of an analogue meter reading, especially where this involves interpolation between scale points. Electrical noise can also be a source of random errors, as can environmental temperature changes, dust, friction, vibration and similar sources.

The tendency of random errors towards a zero mean suggests that they can be avoided if the same measurement is made a number of times and an average taken. However, this only works if the errors are truly random. In the case of human-operated meter readings, a different value will be recorded on each observation of the meter. If the meter is being read properly from directly above, then the errors will probably be random and an averaging process will produce a very accurate reading. Frequently, though, a human observer will persistently read a meter from one side. The parallax-induced error caused by this is a systematic error rather than a random one, and averaging over any number of readings will not eliminate it. In a similar way, errors due to temperature fluctuations will not be random if there is a net change in the temperature over the period of the readings rather than positive and negative temperature variations about a constant temperature value.

Systematic errors describe errors in the output readings of a measurement system which are unlikely to be revealed by repeated readings, and these therefore cause greater problems. Two major sources of systematic errors are system disturbance during measurement and the effect of modifying inputs, as discussed in Sections 5.2 and 5.3. Other sources of systematic error include bent meter needles, the use of uncalibrated instruments, poor cabling practices and the generation of thermal e.m.f.'s. The latter two sources are considered in Section 5.4. Even when systematic errors due to the above factors have been reduced or eliminated, some errors still remain which are inherent in the manufacture of an instrument.

The *instrument data sheet* supplied by the instrument's manufacturer usually includes a figure which indicates the instrument's accuracy. The accuracy figure normally quoted takes account of the variations in output likely to occur due to factors inherent in the instrument's manufacture (systematic errors in the form of a bias or sensitivity change in the output) and also variations due to random errors. Because random errors are taken account of, the quoted accuracy figure normally has confidence limits of $\pm 2\sigma$ (where σ is the standard deviation – see Chapter 7 for further explanation). These confidence limits are implied by the accuracy figure even if not actually stated. This means that if an inaccuracy of ± 1 per cent is quoted for an instrument, we can assume that 95 per cent of the readings taken have a maximum error of ± 1 per cent of the reading. This probabilistic nature of measurement is allowed for in calibration procedures which also only require measurements to be within documented 95 per cent statistical condfidence limits.

5.2 System disturbance due to measurement

If we were to start with a beaker of hot water and wished to measure its temperature with a mercury-in-glass thermometer, then we should take the thermometer, which would be initially at room temperature, and plunge it into the water. In so doing, we would be introducing a relatively cold mass (the thermometer) into the hot water and a heat transfer would take place between the water and the thermometer. This heat transfer would lower the temperature of the water. While in this case the reduction in temperature would be so small as to be undetectable by the limited measurement resolution of such a thermometer, the effect is finite and clearly establishes the principle that, in nearly all measurement situations, the process of measurement disturbs the system and alters the values of the physical quantities being measured.

Another example is that of measuring car tyre pressures with the type of pressure gauge commonly obtainable from car accessory shops. Measurement is made by pushing one end of the pressure gauge onto the valve of the tyre and reading the displacement of the other end of the gauge against a scale. As the gauge is used, a quantity of air flows from the tyre into the gauge. This air does not subsequently flow back into the tyre after measurement, and so the tyre has been disturbed and the air pressure inside it has been permanently reduced.

Disturbance of electrical circuits during measurement is particularly important. For instance, whenever a voltmeter is applied to measure voltage in a circuit, the voltmeter draws a current and thus loads the circuit. Mechanisms for quantifying this effect, using Thevenin's theorem, are covered at length in Appendix 4.

Thus, as a general rule, the process of measurement always disturbs the system being measured. The magnitude of the disturbance varies from one measurement system to the next and is affected particularly by the type of instrument used for measurement.

Ways of minimizing disturbance of measured systems are important considerations in instrument design. However, a prerequisite for this is an accurate understanding of the mechanisms of system disturbance. In the case of voltage measurement, it is shown in Appendix 4 that the loading effect on the circuit being measured is minimized when the resistance of the measuring instrument is large. Thus the design strategy should be to make the resistance of the voltmeter used as high as possible.

It is an interesting design exercise to consider how this might be achieved if the voltmeter used were a moving coil voltmeter. Such an instrument consists of a coil carrying a pointer mounted in a fixed magnetic field. As current flows through the coil, the interaction between the field generated and the fixed field causes the pointer attached to the coil to turn in proportion to the applied current.

What practical steps can be taken in the design of such an instrument to achieve a high internal resistance? The simplest ways of increasing the input impedance (the resistance) of the meter are either to increase the number of turns in the coil or to construct the same number of coil turns with a higher-resistance material. However, either of these solutions decreases the current flowing in the coil, giving less magnetic torque and thus decreasing the measurement sensitivity of the instrument (i.e. for a given applied voltage, we get less deflection of the pointer). This problem can be overcome by changing the spring constant of the restraining springs of the instrument, such that less torque is required to turn the pointer by a given amount. This, however, reduces the ruggedness of the instrument and also demands better pivot design to reduce friction, and highlights a very important but tiresome principle in instrument design: any attempt to improve the performance of an instrument in one aspect generally decreases the performance in some other aspect. This is an inescapable fact of life with passive instruments such as the type of voltmeter mentioned, and is often the reason for the use of alternative active instruments such as digital voltmeters, where the inclusion of auxiliary power improves performance considerably.

5.3 Modifying inputs in measurement systems

The fact that the static and dynamic characteristics of measuring instruments are specified for particular environmental conditions of temperature and pressure, etc., has already been discussed at considerable length. These specified conditions must be reproduced as closely as possible during calibration exercises because, away from the specified calibration conditions, the characteristics of measuring instruments vary to some extent. The magnitude of this variation is quantified by the two constants known as the sensitivity drift and the zero drift, both of which are generally included in the published specifications for an instrument. Such variations of environmental conditions away from the calibration conditions are described as modifying inputs to the system. The environmental variation is described as a measurement system input because the effect on the system output is the same as if the value of the measured quantity had changed by a certain amount. In any general measurement situation, it is very difficult to avoid modifying inputs, because it is either impractical or impossible to control the environmental conditions surrounding the measurement system. System designers are therefore charged with the task of either reducing the susceptibility of measuring instruments to modifying inputs or alternatively quantifying the effect of modifying inputs and correcting for them in the instrument output reading.

The techniques used to deal with modifying inputs and minimize their effect on the final output measurement follow a number of routes as discussed below.

Careful instrument design

Careful instrument design is the most useful weapon in the battle against modifying inputs, by reducing the sensitivity of an instrument to modifying inputs to as low a level as possible. In the design of strain gauges, for instance, the element should be constructed from a material whose resistance has a very low temperature coefficient (i.e. the variation of the resistance with temperature is very small). For many instruments, however, it is not possible to reduce their sensitivity to modifying inputs to a satisfactory level by simple design adjustments, and other measures have to be taken.

Method of opposing inputs

The method of opposing inputs compensates for the effect of a modifying input in a measurement system by introducing an equal and opposite modifying input which cancels it out. One example of how this technique is applied is in the type of millivoltmeter shown in Figure 5.1. This consists of a coil suspended in a fixed magnetic field produced by a permanent magnet. When an unknown voltage is applied to the coil, the magnetic field due to the current interacts with the fixed field and causes the coil (and a pointer attached to the coil) to turn. If the coil resistance is sensitive to temperature, then any modifying input to the system in the form of a temperature change will alter the value of the coil current for a given applied voltage and so alter the pointer output reading. Compensation for this is made by introducing a compensating resistance R_{comp} into the circuit, where R_{comp} has a temperature coefficient which is equal in magnitude but opposite in sign to that of the coil.

High-gain feedback

The benefit of adding high-gain feedback to many measurement systems is illustrated by considering the case of the voltage measuring instrument whose block diagram is shown in Figure 5.2. In this system, the unknown voltage E_i is applied to a coil of torque constant K_c, and the torque induced turns a pointer

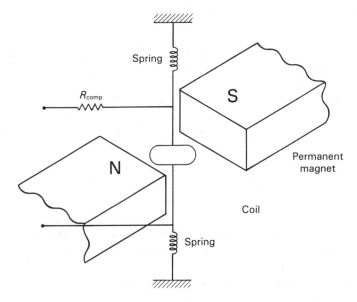

Figure 5.1 Millivoltmeter

against the restraining action of a spring with spring constant K_s. The effect of modifying inputs on the coil and spring constants are represented by variables D_c and D_s. The displacement of the pointer X_o is given by:

$$X_o = K_m \cdot K_s \cdot E_i$$

in the absence of modifying inputs.

In the presence of modifying inputs, however, both K_c and K_s change and the relationship between X_o and E_i can be affected greatly. It therefore becomes difficult or impossible to calculate E_i from the measured value of X_o.

Consider now what happens if the system is converted into a high-gain, closed-loop one, as shown in Figure 5.3, by adding an amplifier of gain constant

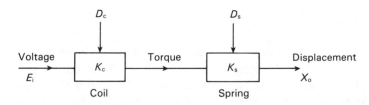

Figure 5.2 Block diagram for voltage measuring instrument

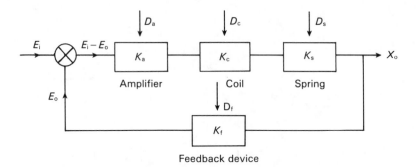

Figure 5.3 Block diagram of voltage measuring instrument with high-gain feedback

K_a and a feedback device with gain constant K_f. Assume also that the effect of modifying inputs on the values of K_a and K_f are represented by D_a and D_f. The feedback device feeds back a voltage E_o proportional to the pointer displacement X_o. This is compared with the unknown voltage E_i by a comparator and the error is amplified.

Writing down the equations of the system, we have

$$E_o = K_f \cdot X_o$$

$$X_o = (E_i - E_o) \cdot K_a \cdot K_c \cdot K_s = (E_i - K_f \cdot X_o) \cdot K_a \cdot K_c \cdot K_s$$

thus:

$$E_i \cdot K_a \cdot K_c \cdot K_s = (1 + K_f \cdot K_a \cdot K_c \cdot K_s) \cdot X_o$$

i.e.

$$X_o = \frac{K_a \cdot K_c \cdot K_s}{1 + K_f \cdot K_a \cdot K_c \cdot K_s} \cdot E_i \qquad (5.1)$$

Because K_a is very large (it is a high gain amplifier): $K_f \cdot K_a \cdot K_c \cdot K_s \gg 1$, and equation (5.1) reduces to

$$X_o = \frac{E_i}{K_f}$$

This is a highly important result because we have reduced the relationship between X_o and E_i to one which involves only K_f. The sensitivity of the gain constants K_a, K_c and K_s to the modifying inputs D_a, D_c and D_s has thereby been rendered irrelevant and we only have to be concerned with one modifying input D_f. Conveniently, it is usually an easy matter to design a feedback device which is insensitive to modifying inputs; this is much easier than trying to make a coil or spring insensitive. Thus high-gain feedback techniques are often a very effective way of reducing a measurement system's sensitivity to modifying inputs. One potential problem which must be mentioned, however, is that there is a possibility that high-gain feedback will cause instability in the system. Any application of this method must therefore include careful stability analysis of the system.

Signal filtering

One frequent problem in measurement systems is corruption of the output reading by periodic noise, often at a frequency of 50 Hz caused by pick-up through the close proximity of the measurement system to apparatus or current-carrying cables operating on a mains supply. Periodic noise corruption at higher frequencies is also often introduced by mechanical oscillation or vibration within some component of a measurement system. The amplitude of all such

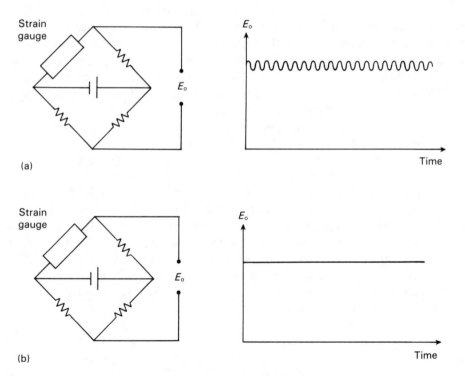

Figure 5.4: (a) noise-corrupted output of bridge circuit measuring resistance of strain gauge; (b) effect of adding low-pass filter

noise components can be substantially attenuated by the inclusion of filtering of an appropriate form in the system. Band-stop filters can be especially useful where corruption is of one particular known frequency, or, more generally, low-pass filters are employed to attenuate all noise in the range of 50 Hz and higher frequencies. The subject of signal filtering is considered at greater length in Chapter 6.

Measurement systems with a low-level output, such as a bridge circuit measuring a strain-gauge resistance, are particularly prone to noise, and Figure 5.4(a) shows the typical corruption of a bridge output by 50 Hz pick-up. The beneficial effect of putting a simple passive RC low-pass filter across the output is shown in Figure 5.4(b).

5.4 Other sources of error

Connecting leads

In connecting together the components of a measurement system, a common source of error is the failure to take proper account of the resistance of connecting leads (or pipes in the case of pneumatically or hydraulically actuated measurement systems). In typical applications of a resistance thermometer, for instance, it is common to find the thermometer separated from other parts of the measurement system by perhaps 30 m. The resistance of such a length of 7/0.0076 copper wire is 2.5Ω and there is a further complication that such wire has a temperature coefficient of $1 m\Omega/^{\circ}C$.

Careful consideration thus needs to be given to the choice of connecting leads. Not only should they be of adequate cross-section so that their resistance is minimized, but they should be adequately screened if they are thought likely to be subject to electrical or magnetic fields which could otherwise cause induced noise. Where screening is thought essential, then the routing of cables also needs careful planning. In one application in the author's personal experience, involving instrumentation of an electric-arc steelmaking furnace, screened signal-carrying cables between transducers on the arc furnace and a control room at the side of the furnace were initially corrupted by high-amplitude 50 Hz noise. However, by changing the route of the cables between the transducers and the control room, the magnitude of this induced noise was reduced by a factor of about ten.

Thermal e.m.f.'s

Whenever metals of two different types are connected together, a thermal e.m.f. is generated according to the temperature of the joint. This is the physical principle on which temperature-measuring thermocouples operate. Such thermal e.m.f.'s are only a few millivolts in magnitude and so this effect is only significant when typical voltage output signals of a measurement system are of a similar low magnitude.

One such situation is where one e.m.f. measuring instrument is used to monitor the output of several thermocouples measuring the temperatures at different points in a process control system. This requires a means of automatically switching the output of each thermocouple to the measuring instrument in turn. Nickel–iron reed-relays with copper connecting leads are commonly used to provide this switching function. This introduces a thermocouple effect of magnitude $40 \, \mu V/^\circ C$ between the reed-relay and the copper connecting leads. There is no problem if both ends of the reed-relay are at the same temperature because then the thermal e.m.f.'s will be equal and opposite and so cancel out. There are several recorded instances, however, where, because of lack of awareness of the problem, poor design has resulted in the two ends of a reed-relay being at different temperatures and resulting in a net thermal e.m.f. The serious error that this introduces is clear. For a temperature difference between the two ends of only $2 \, ^\circ C$, the thermal e.m.f. is $80 \, \mu V$, which is very large compared with a typical thermocouple output level of $400 \, \mu V$.

Another example of the difficulties that thermal e.m.f.'s can create becomes apparent in considering the following problem which was reported in a current measuring system. This had been designed such that the current flowing in a particular part of a circuit was calculated by applying it to an accurately calibrated wire-wound resistance of value 100Ω and measuring the voltage drop across the resistance. In calibration of the system, a known current of $20 \, \mu A$ was applied to the resistance and a voltage of $2.20 \, mV$ was measured by an accurate high-impedance instrument. Simple application of Ohm's law reveals that such a voltage reading indicates a current value of $22 \, mA$. What then was the explanation for this discrepancy? The answer, once again, is a thermal e.m.f. Because the designer was not aware of thermal e.m.f.'s, the circuit had been constructed such that one side of the standard resistance was close to a power transistor, creating a difference in temperature between the two ends of the resistor of $2 \, ^\circ C$. The thermal e.m.f. associated with this was sufficient to account for the 10 per cent measurement error found.

5.5 Error reduction using intelligent instruments

Intelligence in instruments can bring about a gross reduction in the level of output error in measurement systems which are subject to the types of error discussed earlier in this chapter. The ability of intelligent instruments to achieve this does, however, require that the following pre-conditions be satisfied:

1. The physical mechanism by which a measurement transducer is affected by ambient condition changes must be fully understood and all physical quantities which affect the transducer output must be identified.
2. The effect of each ambient variable on the output characteristic of the measurement transducer must be quantified.
3. Suitable secondary transducers for monitoring the value of all relevant ambient variables must be available for input to the intelligent instrument.

Condition 1 above means that the thermal expansion/contraction of all elements within a transducer must be considered in order to evaluate how it will respond to ambient temperature changes. Similarly, the transducer response, if any, to changes in ambient pressure, humidity, gravitational force or power supply level (active instruments) must be examined.

Quantification of the effect of each ambient variable on the characteristics of the measurement transducer is then necessary, as stated in condition 2. Analytic quantification of ambient condition changes from purely theoretical consideration of the construction of a transducer is usually extremely complex and so is normally avoided. Instead, the effect is quantified empirically in laboratory tests where the output characteristic of the transducer is observed as the ambient environmental conditions are changed in a controlled manner.

Once the ambient variables affecting a measurement transducer have been identified and their effect quantified, an intelligent instrument can be designed which includes secondary transducers to monitor the value of the ambient variables. Suitable transducers which will operate satisfactorily within the environmental conditions prevailing for the measurement situation must of course exist, as stated in condition 3.

Application of intelligent instruments to reduce random errors

Many cases of random error in measurement systems arise because of human errors in observing meters, and intelligent instruments can make no contribution towards reducing these. However, some sources of random error arise from the presence of electrical noise or mechanical vibrations which affect the measurement system. In such cases, intelligent instruments are programmed to take the same measurement a number of times within a short space of time and perform simple averaging on the readings before displaying an output measurement.

Application of intelligent instruments to reduce systematic errors

The contribution of intelligent instruments in the reduction of sequential errors is subject to the three conditions already stated, and the degree of applicability varies according to the features of each particular measurement situation. Each measurement situation must therefore be considered separately.

In the case of electrical circuits in which the system is disturbed by the loading effect of the measuring instrument, an intelligent instrument can readily correct for measurement errors by applying equations such as (A4.2) with the resistance of the measuring instrument inserted. Similarly for measurement situations containing bridge circuits, equation (A4.7) can be manipulated into a form where R_u is expressed as a function of V_i, V_m and the other circuit resistances if the resistance of the measuring instrument R_m is known. Then an intelligent instrument can use that expression to compute the unknown value of R_u.

Intelligent instruments are particularly useful for dealing with measurement errors due to modifying inputs. Secondary transducers are provided within the instrument to monitor the magnitude of environmental conditions such as temperature and pressure which can affect the characteristics of the primary measurement transducer. The computer within the instrument then corrects the measurement obtained from the primary transducer according to the values read by the secondary transducers. This presupposes that all the modifying inputs in a measurement situation have been correctly identified and quantified, and that suitable secondary transducers exist to monitor these modifying inputs.

Suitable care must always be taken when introducing a microcomputer into a measurement system to avoid creating new sources of measurement noise. This is particularly so where one microcomputer is used to process the output of several transducers and is connected to them by signal wires. In such circumstances, the connections and connecting wires can create noise through electrochemical potentials, thermoelectric potentials, offset voltages introduced by common mode impedances, and ac noise at power, audio and radio frequencies. Recognition of all these possible noise sources allows them to be eliminated in most cases by employing good measurement system construction practices. All remaining noise sources are usually eliminated by the provision of a set of four earthing circuits within the interface fulfilling the following functions:

- Power earth: provides a path for fault currents due to power faults.

- Logic earth: provides a common line for all logic circuit potentials.
- Analogue earth (ground): provides a common reference for all analogue signals.
- Safety earth: connected to all metal parts of equipment to protect personnel should power lines come into contact with metal enclosures.

6

Processing of measurement signals

The last chapter was concerned with identifying the major sources of measurement errors and designing measurement systems such that errors are minimized. However, it is not always possible, or at least it is not always cost-effective, to remove all measurement errors. In this circumstance, where the measurement signals are corrupted by errors, much can be done to improve the quality of measurement data by appropriate processing of the measurement signals.

Signal processing is thus concerned with improving the quality of the reading or signal at the output of a measurement system. The form which signal processing takes depends on the nature of the raw output signals from the measurement transducers. Procedures of signal amplification, signal attenuation, signal linearization, bias removal and signal filtering are all particular forms of signal processing which are applied according to the form of correction required in each raw signal.

The implementation of signal processing procedures can be carried out either by analogue techniques or by digital computation. For the purposes of explaining the procedures involved, this chapter concentrates on analogue signal processing and concludes with a relatively brief discussion of the equivalent digital signal processing techniques. One particular reason for this method of treatment is that some prior analogue signal conditioning is often necessary, even when the major part of the signal processing is carried out digitally.

Choice between analogue and digital signal processing is largely determined by the degree of accuracy required in the signal processing procedure. Digital processing is very much more accurate than analogue processing but the cost of the equipment involved is much greater and the processing time is much longer. Normal practice, therefore, is to use analogue techniques for all signal processing tasks except where the accuracy of this is insufficient. It should be noted also that in some measurement situations where a physical quantity is measured by an inherently inaccurate transducer, the extra accuracy provided by digital signal processing is insignificant and therefore such expensive and slow processing techniques are inappropriate.

6.1 Signal amplification

Signal amplification is carried out when the typical signal output level of a measurement transducer is considered to be too low. Amplification by analogue means is carried out by an operational amplifier. This is normally required to have a high-input impedance so that its loading effect on the transducer output signal is minimized. In some circumstances, such as when amplifying the output signal from accelerometers and some optical detectors, the amplifier must also have a high-frequency response, to avoid distortion of the measured quantity–output reading relationship.

The operational amplifier is an electronic device which has two input terminals and one output terminal, the two inputs being known as the inverting input and non-inverting input respectively. When used for signal processing duties, it is connected in the configuration shown in Figure 6.1. The raw (unprocessed) signal V_i is connected to the inverting input through a resistor R_1 and the non-inverting input is connected to ground. A feedback path is provided from the output terminal through a resistor R_2 to the inverting input terminal. The processed signal V_o at the output terminal is then related to the voltage V_i at the input terminal by the expression (assuming ideal operational amplifier characteristics)

$$V_o = -\frac{R_2}{R_1} \cdot V_i \qquad (6.1)$$

The amount of signal amplification is therefore defined by the relative values of R_1 and R_2. This ratio between R_1 and R_2 in the amplifier configuration is often known as the amplifier gain. If, for instance, $R_1 = 1$ MΩ and $R_2 = 10$ MΩ, an amplification factor of 10 is obtained (i.e. gain = 10). It is important to note

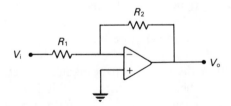

Figure 6.1 Operational amplifier connected for signal amplification

that in this standard way of connecting the operational amplifier, the sign of the processed signal is inverted. This can be corrected for, if necessary, by feeding the signal through a further amplifier set up for unity gain ($R_1 = R_2$). This inverts the signal again and returns it to its original sign.

6.2 Signal attenuation

One method of attenuating signals by analogue means is to use a potentiometer connected in a voltage-dividing circuit, as shown in Figure 6.2. For the potentiometer wiper positioned a distance X_w along the resistance element of total length X_t, the voltage level of the processed signal V_o is related to the voltage level of the raw signal V_i by the expression:

$$V_o = \frac{X_w}{X_t} \cdot V_i$$

An alternative device to the potentiometer for signal attenuation is the operational amplifier. This is connected in exactly the same way as for an amplifier as shown in Figure 6.1, but R_1 is chosen to be greater than R_2. Equation (6.1) still holds and therefore if R_1 is chosen to be 10 MΩ and R_2 as 1 MΩ, an attenuation factor of 10 is achieved (gain = 0.1). Use of an operational amplifier as an attenuating device is a more expensive solution than using a potentiometer, but it has advantages in terms of its smaller size and low power consumption.

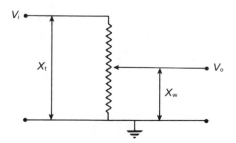

Figure 6.2 Potentiometer in voltage dividing circuit

6.3 Signal linearization

Several types of transducer used in measuring instruments have an output which is a non-linear function of the measured quantity input. In many cases, this

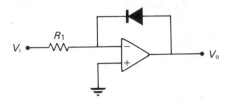

Figure 6.3 Operational amplifier connected for signal linearization

non-linear signal can be converted to a linear one by special operational amplifier configurations which have an equal and opposite non-linear relationship between the amplifier input and output terminals.

For example, light intensity transducers typically have an exponential relationship between the output signal and the input light intensity, i.e.

$$V_o = K \cdot \exp(-\alpha Q) \qquad (6.2)$$

where Q is the light intensity, V_o is the voltage level of the output signal, and K and α are constants.

If a diode is placed in the feedback path between the input and output terminals of the amplifier, as shown in Figure 6.3, the relationship between the amplifier output voltage V_2 and input voltage V_1 is given by

$$V_2 = C \cdot \log_e(V_1) \qquad (6.3)$$

If the output of the light transducer with characteristic given by equation (6.2) is conditioned by an amplifier of characteristic given by equation (6.3), the voltage level of the processed signal is given by

$$V_2 = C \cdot \log(K) - \alpha \cdot C \cdot Q \qquad (6.4)$$

Expression (6.4) shows that the output signal now varies linearly with light intensity Q but with an offset of $C \cdot \log(K)$. This offset would normally be removed by further signal conditioning, as described below.

6.4 Bias removal

Sometimes, either because of the nature of the measurement transducer itself, or as a result of other signal conditioning operations (see Section 6.3), a bias exists in the output signal. This can be expressed mathematically for a physical

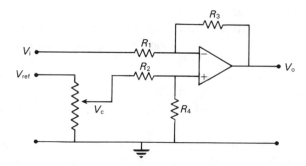

Figure 6.4 Operational amplifier connected in differential amplification mode

quantity x and measurement signal y as

$$y = K \cdot x + C \tag{6.5}$$

where C represents a bias in the output signal which needs to be removed by signal processing. Analogue processing consists of using an operational amplifier connected in a differential amplification mode, as shown in Figure 6.4. Referring to this circuit, for $R_1 = R_2$ and $R_3 = R_4$, the output V_o is given by

$$V_o = (R_3/R_1)(V_c - V_i) \tag{6.6}$$

where V_i is the unprocessed signal y equal to $(K \cdot x + C)$ and V_c is the output voltage from a potentiometer supplied by a known reference voltage V_{ref}, which is set such that $V_c = C$. Now, substituting these values for V_i and V_c into equation (6.6) and referring the quantities back into equation (6.5), we find that

$$y = K' \cdot x \tag{6.7}$$

where the new constant K' is related to K by the amplifier gain factor R_3/R_1. It is clear that a straight-line relationship now exists between the measurement signal y and the measured quantity x and that the unwanted bias has been removed.

6.5 Signal filtering

Signal filtering consists of processing a signal to remove a certain band of frequencies within it. The band of frequencies removed can be either at the low-frequency end of the frequency spectrum, at the high-frequency end, at both ends, or in the middle of the spectrum. Filters to perform each of these operations are known, respectively, as low-pass filters, high-pass filters, band-pass filters and band-stop filters. All such filtering operations can be carried out by either analogue or digital methods.

The result of filtering can be readily understood if the analogy with a procedure such as sieving soil particles is considered. Suppose that a sample of soil A is passed through a system of two sieves of differing meshes such that the soil is divided into three parts, B, C and D, consisting of large, medium and small particles, as shown in Figure 6.5. Suppose that the system also has a mechanism for delivering one or more of the separated parts, B, C and D as the system output. If the graded soil output consists of parts C and D, the system is behaving as a low-pass filter (rejecting large particles), whereas if it consists of parts B and C, the system is behaving as a high-pass filter (rejecting small particles). Other options are to deliver just part C (band-pass filter mode) or parts B and D together (band-stop filter mode). As any gardener knows, however, such perfect sieving is not achieved in practice and any form of graded soil output always contains a few particles of the wrong size.

Signal filtering consists of selectively passing or rejecting low-, medium-

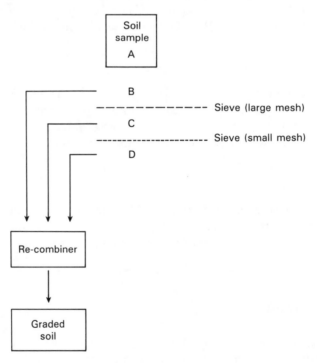

Figure 6.5 Soil sieving analogy of signal filtering

and high-frequency signals from the frequency spectrum of a general signal. The range of frequencies passed by a filter is known as the *pass band*, the range not passed is known as the *stop band*, and the boundary between the two ranges is known as the *cut-off frequency*. To illustrate this, consider a signal whose frequency spectrum is such that all frequency components in the frequency range from zero to infinity have equal magnitude. If this signal is applied to an ideal filter, then the outputs for a low-pass filter, high-pass filter, band-pass filter and band-stop filter, respectively, are shown in Figure 6.6. Note that for the latter two types, the bands are defined by a pair of frequencies rather than by a single cut-off frequency.

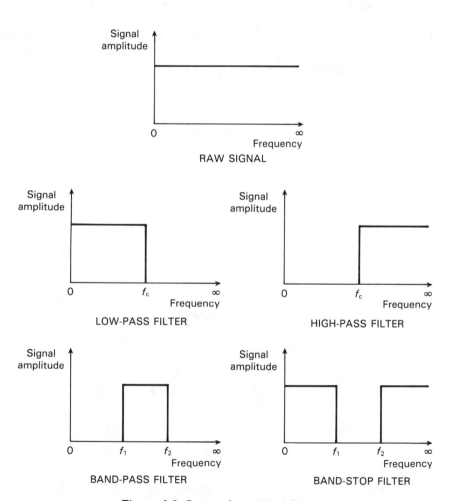

Figure 6.6 Output from ideal filters

Just as in the case of the soil sieving analogy presented above, the signal filtering mechanism is not perfect, with unwanted frequency components not being erased completely but only attenuated by varying degrees instead, i.e. the filtered signal always retains some components (of a relatively low magnitude) in the unwanted frequency range. There is also a small amount of attenuation of frequencies within the pass band which increases as the cut-off frequency is approached. Figure 6.7 shows the typical output characteristics of a practical constant-k* filter designed respectively, for high-pass, low-pass, band-pass and band-stop filtering. Filter design is concerned with trying to obtain frequency rejection characteristics which are as close to the ideal as possible. However, improvement in characteristics is only achieved at the expense of greater complexity in the design. The filter chosen for any given situation is therefore a compromise between performance, complexity and cost.

In the majority of measurement situations, the physical quantity being measured has a value which is either constant or only changing slowly with time. In these circumstances, the most common types of signal corruption are

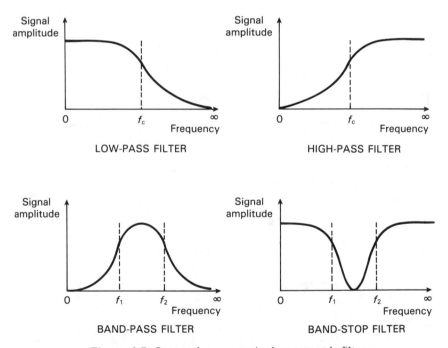

Figure 6.7 Output from practical constant-k filters

* 'constant-k' is a term used to describe a common class of passive filters, as discussed in the following section.

high-frequency noise components, and the type of signal processing element required is a low-pass filter. In a few cases, the measured signal itself has a high frequency, for instance when mechanical vibrations are being monitored, and the signal processing required is the application of a high-pass filter to attenuate low-frequency noise components. Band-stop filters can be used where a measurement signal is corrupted by noise at a particular frequency. Such noise is frequently due to mechanical vibrations or proximity of the measurement circuit to other electrical apparatus.

The design of analogue filters is considered in detail in Appendix 5. A discussion on equivalent digital filters is included in the next section.

6.6 Digital signal processing

Digital techniques achieve much greater levels of accuracy in signal processing than equivalent analogue methods. However, the time taken to process a signal digitally is much longer than that required to carry out the same operation by analogue techniques, and the equipment required is more expensive. Some care is thus needed in making the correct choice between digital and analogue methods in a particular signal processing application.

While digital signal processing elements in a measurement system can exist as separate units, it is more usual to find them as an integral part of an intelligent instrument. However, their construction and mode of operation are the same irrespective of whether they exist physically as separate boxes or are part of an intelligent instrument.

The hardware aspect of a digital signal processing element consists of a digital computer and analogue interface boards. The actual form which signal processing takes depends on the software program executed by the processor. Before consideration is given to this, however, some theoretical aspects of signal sampling need to be discussed.

As mentioned earlier, digital computers require signals to be in digital form whereas most instrumentation transducers have an output signal in analogue form. Analogue-to-digital conversion is therefore required at the interface between analogue transducers and the digital computer. The procedure followed is to sample the analogue signal at a particular moment in time and then convert the analogue value to an equivalent digital one. This conversion takes a certain finite time, during which the analogue signal can be changing in value. The next sample of the analogue signal cannot be taken until the conversion of the last sample to digital form is completed. The representation within a digital computer of a continuous analogue signal is therefore a sequence of samples whose pattern only approximately follows the shape of the original signal. This pattern of samples taken at successive, equal intervals of

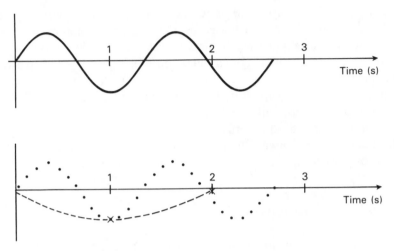

Figure 6.8 Conversion of continuous analogue signal to discrete sampled signal

time is known as a discrete signal. The process of conversion between a continuous analogue signal and a discrete digital one is illustrated for a sine wave in Figure 6.8.

The raw analogue signal in Figure 6.8 has a frequency of approximately 0.75 cycles/s. With the rate of sampling shown, which is approximately 11 samples/s, reconstruction of the samples matches the original analogue signal very well. If the rate of sampling were decreased, the fit between the reconstructed samples and the original signal would be less good. If the rate of sampling was very much less than the frequency of the raw analogue signal, such as 1 sample/s, only the samples marked ×　in Figure 6.8 would be obtained. Fitting a line through these ×s incorrectly estimates a signal whose frequency is approximately 0.25 cycles/s. This phenomenon, whereby the process of sampling transmutes a high-frequency signal into a lower frequency one, is known as *aliasing*. To avoid aliasing, it is necessary theoretically for the sampling rate to be at least twice the highest frequency in the analogue signal sampled. In practice, sampling rates of between five and ten times the highest frequency signal are normally chosen so that the discrete sampled signal is a close approximation to the original analogue signal in amplitude as well as frequency.

Problems can arise in sampling when the raw analogue signal is corrupted by high-frequency noise of unknown characteristics. It would be normal practice to choose the sampling interval as, say, a ten-times multiple of the frequency of the measurement component in the raw signal. If such a sampling interval is chosen, aliasing can in certain circumstances transmute high-frequency noise components into the same frequency range as the measurement component in the signal, thus giving erroneous results. This is one of the circumstances mentioned earlier, where prior analogue signal conditioning in the form of a low-pass filter must be carried out before processing the signal digitally.

One further factor which affects the quality of a signal when it is converted from analogue to digital form is *quantization*, which describes the procedure whereby the continuous analogue signal is converted into a number of discrete levels. At any particular value of the analogue signal, the digital representation is either the discrete level immediately above this value or the discrete level immediately below it. If the difference between two successive discrete levels is represented by the parameter Q, then the maximum error in each digital sample of the raw analogue signal is $\pm Q/2$. This error is known as the quantization error and is clearly proportional to the resolution of the analogue-to-digital converter, i.e. to the number of bits used to represent the samples in digital form.

Once a satisfactory digital representation, in discrete form, of an analogue signal has been obtained, the procedures of signal amplification, signal attenuation and bias removal become trivial. For signal amplification and attenuation, all samples have to be multiplied or divided by a fixed constant. Bias removal involves simply adding or subtracting a fixed constant from each sample of the signal.

Signal linearization requires *a priori* knowledge of the type of non-linearity involved, in the form of a mathematical equation which expresses the relationship between the output measurements from an instrument and the value of the physical quantity being measured. This can be obtained either theoretically through knowledge of the physical laws governing the system or empirically using input–output data obtained from the measurement system under controlled conditions. Once this relationship has been obtained, it is used to calculate the value of the measured physical quantity corresponding to each discrete sample of the measurement signal. While the amount of computation involved in this is greater than for the trivial cases of signal amplification, etc., already mentioned, the computational burden is still relatively small in most measurement situations.

Digital signal processing can also perform all the filtering functions mentioned earlier in respect of analogue filters, i.e. low-pass, high-pass, band-pass and band-stop. However, the design of digital filters requires a level of theoretical knowledge, including the use of z-transform theory, which is outside the range of this book. The reader interested in digital filter design is therefore referred to specialist texts on the subject.[1,2]

7

Analysis and presentation of measurement data

The various sources of error which might occur in a measurement system, and the procedures which can be followed to reduce or avoid them, were covered at length in Chapter 5. The last chapter continued by discussing ways of reducing still further the errors present at the output of measurement systems by appropriate processing of the measurement signals. If the various measurement procedures and signal processing functions recommended in these two chapters are applied, then the measurement system designer will have done all that is reasonably possible to minimize the magnitude of errors in measurements.

What remains to be discussed is how to quantify the total error present at the output of a measurement system, combining as necessary the separate error levels at the output of individual instruments which are part of the measurement system. A further subject of importance is consideration of the optimum means of presenting measurement data for its use in the execution of quality control functions.

This chapter is therefore concerned with these two subjects, i.e., how to analyse the set of data which forms the output of a measurement system in order to extract the maximum amount of information from it, and then how to present the measurement data in the most useful form.

To achieve these objectives, quantification of each separate systematic and random error is required first. The means of quantifying systematic errors has already been presented in Chapter 5, but in the case of random errors, statistical techniques are often needed, as discussed in Section 7.1. Following this, the cumulative effect of all errors on overall measurement system accuracy can be calculated, as indicated in Section 7.2. Finally, in Section 7.3, the various available ways of presenting measurement data are discussed. Formal definitions of many of the statistical terms used in this chapter can be found in references 1 and 2.

7.1 Statistical analysis of data

Concepts of probability

Before proceeding further with the quantification of random errors, it is necessary to introduce some concepts of probability. Any event has a probability of occurrence, which can be quantified on a scale of 0 to 1 or 0 per cent to 100 per cent. If a coin is tossed, then the probability of it coming down heads is 0.5 or 50 per cent. If a six-sided die is thrown once, the probability of throwing a 6 is 1 in 6, i.e. 16.67 per cent. In these simple examples, all possible outcomes are unbiassed, i.e. they are all equally likely to happen. The probability of a particular outcome A occurring when there are N equally possible outcomes can be expressed mathematically as

$$P(A) = \frac{1}{N} \qquad (7.1)$$

Going a little further, suppose that we throw two dice together. What is the probability that the total of the two numbers obtained is nine? The total number of possible combinations is 36. Four of these combinations (3–6, 4–5, 5–4, 6–3) total 9. Thus the probability of throwing a total of 9 with two dice is 4 in 36 (1 in 9) or 11.1 per cent. The probability that such an outcome B occurs when there are n ways in which B can occur out of a total of N possible outcomes can be expressed mathematically as

$$P(B) = \frac{n}{N} \qquad (7.2)$$

Expression (7.2) is not in fact the best way of expressing probabilities because it implies that if we throw the dice nine times we will get a total score of 9 once. Simple experimentation will disprove this. We might not get a total of 9 at all or we might get it more than once out of the nine throws. In fact, throwing the dice nine times, we are quite likely to get a total of 9 on none, one or two of the throws, which is 0 per cent, 11.1 per cent or 22.2 per cent of the throws. If we increase the sample size by, say, throwing the dice ninety times, we would be extremely unlucky to not score a 9 at all and typically would score a 9 on either nine, ten or eleven of the throws, which is 10.0 per cent, 11.1 per cent or 12.2 per cent. This shows that as we increase the sample size (number of throws), the actual number of times that we score a total of 9 moves closer to 11.1 per cent. Only if we throw the dice an infinite number of times would we be certain that the number of times we scored a 9 would equal 11.1 per cent of the total. Mathematically, it is thus better to define probability as a number that the result tends towards as the sample size tends towards infinity:

$$P(B) = \lim_{n \to \infty} \frac{n}{N} \tag{7.3}$$

Many errors are also random in nature, but not in quite the same way as the examples given above. If a digital temperature-measuring instrument with a measurement resolution of $0.1\,^\circ C$ was used to measure a constant temperature of $100\,^\circ C$ a large number of times, the temperature readings might range from $99.0\,^\circ C$ to $101.0\,^\circ C$. Such variations in the output readings are due to random changes in the measurement system parameters. However, we should expect these random changes to affect the output reading by a large amount much less often than by a small amount (experimentation would confirm this). Thus small errors in the measurement will occur much more frequently than large errors, and therefore the distribution of the measurements between $99\,^\circ C$ and $101\,^\circ C$ will not be uniform. The greatest concentration in output readings will occur within a narrow band close to $100\,^\circ C$, and the density of readings will get progressively thinner towards the extremity temperature readings of $99\,^\circ C$ and $101\,^\circ C$. There will, for example, be many more readings of $99.9\,^\circ C$ than $99.1\,^\circ C$.

In any measurement situation which is subject to random errors, it is standard practice to obtain the measurement by averaging over several readings. The mean of the measurements will only be exactly equal to the value of the measured quantity when the number of measurements is equal to infinity, but it is clearly never possible in practice to take a measurement an infinite number of times. However, the closeness to the true value of the mean of a finite number of measurements can be expressed as a statistical probability.

Mean and median values

When taking the same measurement a number of times in order to minimize random errors, it is preferable to use several different observers if humans are being used to make meter readings, as this ensures as far as possible that systematic, parallax-induced errors are eliminated. The most likely value from this measurement data set can then be expressed as either the mean or the median value.

For a set of n measurements $x_1, x_2, ..., x_n$, the most likely true value is the mean given by

$$x_{\text{mean}} = \frac{x_1 + x_2 + ... + x_n}{n} \tag{7.4}$$

This is valid for all data sets where the measurement errors are distributed equally about the line of zero error, i.e. where the positive errors are balanced in quantity and magnitude by the negative errors.

When the number of values in the data set is large, however, calculation of the mean value is tedious, and it is more convenient to use the median value, this being a close approximation to the mean value. The *median* is given by the middle value when the measurements in the data set are written down in ascending order of magnitude. For a set of n measurements $x_1, x_2, ..., x_n$ written down in ascending order of magnitude, the median value is given by

$$x_{median} = x_{(n+1)/2} \qquad (7.5)$$

Thus, for a set of nine measurements $x_1, x_2, ..., x_9$, the median value is x_5.

For an even number of data values, the median value is mid-way between the centre two values, i.e. for ten measurements $x_1, ..., x_{10}$, the median value is given by

$$\frac{x_5 + x_6}{2}$$

Suppose that, in a particular measurement situation, a mass is measured by a beam-balance, and the following set of readings is obtained at a particular time by different observers:

$$\begin{matrix} 81.6 & 81.1 & 81.4 & 80.9 & 81.1 & 80.5 & 81.3 & 80.8 & 81.2 & 81.8 \\ 81.1 & 81.5 & 81.0 & 81.3 & 81.1 & 80.8 & 81.3 & 81.6 & 81.1 & \end{matrix} \qquad (7.6)$$

The mean value of this set of data is 81.18, calculated according to equation (7.4). The median value is 81.1, which is the middle value if the data values are written down in ascending order, starting at 80.5 and ending at 81.8.

Standard deviation and variance

The probability that the mean or median value of a data set represents the true measurement value depends on how widely scattered the data values are. If the values in the last set of mass measurements (7.6) had ranged from 79 up to 83, our confidence in the mean value would be much less. The spread of values about the mean is analysed by first calculating the deviation of each value from the mean. For any general value x_i, the deviation d_i is given by

$$d_i = x_i - x_{mean}$$

The extent to which n measurement values are spread about the mean can now be expressed by the standard deviation σ, where σ is given by

$$\sigma = \sqrt{\left(\frac{d_1^2 + d_2^2 + \ldots + d_n^2}{n-1}\right)} \tag{7.7}$$

This spread can alternatively be expressed by the variance V, which is the square of the standard deviation, i.e.

$$V = \sigma^2$$

Frequency distributions

A further and very powerful way of analysing the pattern in which measurements deviate from the mean value is to use graphic techniques. The simplest way of doing this is by means of a *histogram*, where various error bands of equal widths are defined and the number of measurements with error within each range recorded. Figure 7.1 shows a histogram drawn from the set of mass measurement data in (7.6). This has the characteristic shape shown by truly random data, with symmetry about the zero error line.

As the number of measurements increases, smaller error bands can be defined for the error histogram, which retains its basic shape but then consists of a larger number of smaller steps on each side of the peak. In the limit, as the number of measurements approaches infinity, the histogram becomes a smooth curve known as a *frequency distribution curve*, as shown in Figure 7.2. The ordinate of this curve is the frequency of occurrence of each error level, $f(E)$, and the abscissa is the error magnitude, E.

If the height of the frequency distribution curve is standardized such that the area under it is unity, then the curve in this form is known as a *probability curve*, and the height $f(E)$ at any particular error magnitude E is known as the *probability density function (pdf)*. The condition that the area under the curve is unity can be expressed mathematically as

$$\int_{-\infty}^{\infty} f(E) \, dE = 1$$

The probability that the error in any one particular measurement lies between two levels E_1 and E_2 can be calculated by measuring the area under the curve contained between two vertical lines drawn through E_1 and E_2, as shown by the

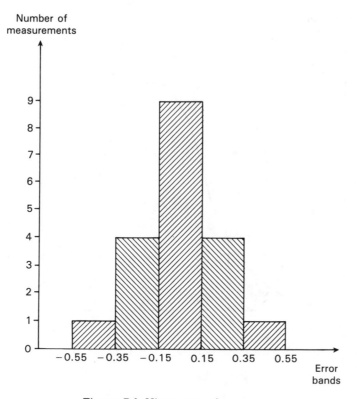

Figure 7.1 Histogram of errors

right-hand hatched area in Figure 7.2. This can be expressed mathematically as

$$P(E_1 < E < E_2) = \int_{E_1}^{E_2} f(E) \, dE \tag{7.8}$$

This expression (7.8) is often known as the *error function*. Of particular importance for assessing the maximum error likely in any one measurement is the *cumulative distribution function (cdf)*. This is defined as the probability of observing a value less than or equal to E_o, and is expressed mathematically as

$$P(E < E_o) = \int_{-\infty}^{E_o} f(E) \, dE \tag{7.9}$$

Thus the cdf is the area under the curve to the left of a vertical line drawn through E_o, as shown by the left-hand hatched area on Figure 7.2.

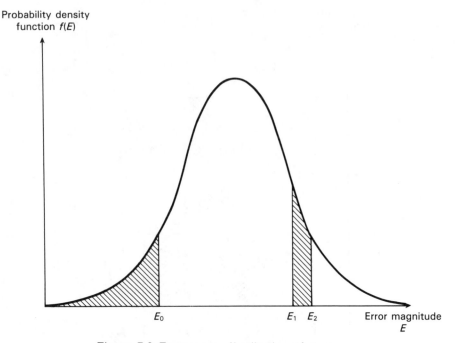

Figure 7.2 Frequency distribution of errors

Three special types of frequency distribution known as the Gaussian, binomial and Poisson distributions exist, and these are very important because most data sets approach closely to one or other of them. The distribution of relevance to data sets containing random measurement errors is the Gaussian distribution.

Gaussian distribution

A Gaussian curve is defined as an error frequency distribution where the error frequency and magnitude are related by the expression

$$f(E) = \frac{1}{\sigma(2\pi)^{1/2}} \exp[-(E-m)^2/2\sigma^2] \tag{7.10}$$

where m is the mean value and the other quantities are as defined before. Most measurement data sets such as the mass values given above in (7.6) fit a Gaussian distribution curve because, if errors are truly random, small deviations from the mean value occur much more often than large deviations, i.e. the number of small errors is much larger than the number of large ones. Alternative names for the Gaussian distribution curve are the *normal distribution* or *bell-shaped distribution*.

The Gaussian distribution curve is symmetrical about the line of zero measurement error, which means that positive errors away from the mean value occur in equal quantities to negative errors in any data set containing measurements subject to random error. If the standard deviation is used as a unit of error, the curve can be used to determine the probability of the error in any particular measurement in a data set being greater than a certain value. By substituting the expression for $f(E)$ in (7.10) into the probability equation (7.8), it can be shown that, for Gaussian-distributed data values, 68 per cent of the values lie within the bounds of $\pm\sigma$. Boundaries of $\pm2\sigma$ contain 95 per cent of data points, and extending the boundaries to $\pm3\sigma$ encompasses 99.7 per cent of data points. The probability of any data point lying outside particular error boundaries can therefore be expressed by the following table.

Error boundaries	Percentage of data points within boundary	Probability of any particular data point being outside boundary
$\pm\sigma$	68.0	32.0
$\pm2\sigma$	95.0	5.0
$\pm3\sigma$	99.7	0.3

Standard error of the mean

The foregoing analysis is only strictly true for measurement sets containing infinite populations. It is not of course possible to obtain an infinite number of data values, and some error must therefore be expected in the calculated mean value of the practical, finite data set available. If several sub-sets are taken from an infinite data population, then, by the *central limit theorem*, the means of the sub-sets will form a Gaussian distribution about the mean of the infinite data set. The error in the mean of a finite data set is usually expressed as the *standard error of the mean*, α, which is calculated as

$$\alpha = \sigma/n^{1/2}$$

This tends towards zero as the number of measurements in the data set is expanded towards infinity. The value obtained from a set of n measurements

$x_1, x_2, ..., x_n$ is then expressed as

$$x = x_{mean} \pm \alpha$$

For the data set of mass measurements in (7.6), $n = 19$, $\sigma = 0.318$ and $\alpha = 0.073$. The mass can therefore be expressed as 81.18 ± 0.07 (for 68 per cent confidence limits).

It is in fact more common to express measurements with 95 per cent confidence limits (2σ boundaries). In this case, $2\sigma = 0.636$, $2\alpha = 0.146$ and the mass can be expressed as 81.18 ± 0.15 (95 per cent confidence limits).

7.2 Estimation of total measurement system errors

A measurement system often consists of several separate components, each of which is subject to systematic and/or random errors. Mechanisms have now been presented for quantifying the errors arising from each of these sources and therefore the total error at the output of each measurement system component can be calculated. What remains to be investigated is how the errors associated with each measurement system component combine together, so that a total error calculation can be made for the complete measurement system.

Error in a product

If the outputs y and z of two measurement system components are multiplied together, the product can be written as

$$P = y \cdot z$$

If the possible error in y is $\pm a \cdot y$ and in z is $\pm b \cdot z$, then the maximum and minimum values possible in P can be written as

$$P_{max} = (y + a \cdot y) \cdot (z + b \cdot z) \qquad ; \qquad P_{min} = (y - a \cdot y) \cdot (z - b \cdot z)$$
$$= y \cdot z + a \cdot y \cdot z + b \cdot y \cdot z + a \cdot y \cdot b \cdot z \qquad\qquad = y \cdot z - a \cdot y \cdot z - b \cdot y \cdot z + a \cdot y \cdot b \cdot z$$

For typical measurement system components with output errors of up to 1 or 2 per cent in magnitude, both a and b are very much less than 1 per cent in magnitude and thus terms in $a \cdot y \cdot b \cdot z$ are negligible compared with the other terms. Therefore we have

$$P_{max} = y \cdot z(1 + a + b) \qquad ; \qquad P_{min} = y \cdot z(1 - a - b)$$

Thus the error in the product P is $\pm(a + b)$.

Example

If the power in a circuit is calculated from measurements of voltage and current in which the calculated maximum errors are, respectively, ± 1 per cent ($a = 1$) and ± 2 per cent ($b = 2$), then the possible error in the calculated power value is ± 3 per cent.

Error in a quotient

If the output measurement y of one system component with possible error $\pm a \cdot y$ is divided by the output measurement z of another system component with possible error $\pm b \cdot z$, then the maximum and minimum possible values for the quotient can be written as

$$Q_{max} = \frac{y + a \cdot y}{z - b \cdot z} \qquad ; \qquad Q_{min} = \frac{y - a \cdot y}{z + b \cdot z}$$

$$= \frac{(y + a \cdot y)(z + b \cdot z)}{(z - b \cdot z)(z + b \cdot z)} \qquad ; \qquad = \frac{(y - a \cdot y)(z - b \cdot z)}{(z + b \cdot z)(z - b \cdot z)}$$

$$= \frac{yz + ayz + byz + abyz}{z^2 - b^2 \cdot z^2} \qquad ; \qquad = \frac{yz - ayz - byz + abyz}{z^2 - b^2 \cdot z^2}$$

For $a \ll 1$ and $b \ll 1$, terms in ab and b^2 are negligible compared with the other terms. Hence

$$Q_{max} = \frac{yz(1 + a + b)}{z^2} \qquad ; \qquad Q_{min} = \frac{yz(1 - a - b)}{z^2}$$

i.e.

$$Q = \frac{y}{z} \pm \frac{y}{z}(a + b)$$

Thus the error in the quotient is $\pm (a + b)$.

Example

If the resistance in a circuit is calculated from measurements of voltage and current where the respective errors are ± 1 per cent ($a = 1$) and ± 0.3 per cent ($b = 0.3$), the likely error in the resistance value is ± 1.3 per cent.

Error in a sum

If the two outputs y and z of separate measurement system components are to be added together, we can write the sum as

$$S = y + z$$

If the maximum errors in y and z are $\pm a \cdot y$ and $\pm b \cdot z$, respectively, we can express the maximum and minimum possible values of S as

$$S_{max} = y + a \cdot y + z + b \cdot z \quad ; \quad S_{min} = y - a \cdot y + z - b \cdot z$$

or

$$S = y + z \pm (a \cdot y + b \cdot z)$$

This relationship for S is not convenient because in this form the error term cannot be expressed as a fraction or percentage of the calculated value for S, and it is also statistically incorrect because the errors in y and z are unlikely to both be at their maximum or minimum values at the same time.

Following statistical analysis, the most probable maximum error in S can be expressed by a quantity e, where e is given by

$$e = [(a \cdot y)^2 + (b \cdot z)^2]^{1/2} \tag{7.11}$$

Thus:

$$S = (y + z) \pm e$$

This can be expressed in the alternative form

$$S = (y + z)(1 \pm f) \tag{7.12}$$

where $f = e/(y + z)$.

Example
A circuit requirement for a resistance of 550Ω is satisfied by connecting together two resistors of nominal values 220Ω and 330Ω in series. If each resistor has a tolerence of ± 2 per cent, the error in the sum calculated according to equations (7.11) and (7.12) is given by

$$e = (0.02 \times 220)^2 + (0.02 \times 330)^2]^{1/2} = 7.93$$

$$f = 7.93/550 = 0.0144$$

Thus the total resistance S can be expressed as

$$S = 550\Omega \pm 7.93\Omega$$

or $S = 550(1 + 0.0144)\Omega$, i.e. $S = 550\Omega \pm 1.4$ per cent.

Error in a difference

If the two outputs y and z of separate measurement systems are to be subtracted from one another, and the possible errors are $\pm a \cdot y$ and $\pm b \cdot z$, then the difference S can be expressed as

$$S = (y - z) \pm e \quad \text{or} \quad S = (y - z)(1 \pm f)$$

where e is calculated as above (equation (7.11)), and $f = e/(y - z)$

Example
A fluid flow rate is calculated from the difference in pressure measured on both sides of an orifice plate. If the pressure measurements are 10.0 bar and 9.5 bar and the error in the pressure measuring instruments is specified as ± 0.1 per cent, then values for e and f can be calculated as

$$e = [(0.001 \times 10)^2 + (0.001 \times 9.5)^2]^{1/2} = 0.0138$$

$$f = 0.0138/0.5 = 0.0276$$

Thus the pressure difference can be expressed as 0.5 bar ± 2.8 per cent. This example illustrates very poignantly the relatively large error which can arise when calculations are made based on the difference between two measurements.

7.3 Presentation of data

We have now covered what the various sources of systematic and random errors are in a measurement system, how the magnitude of each error component is calculated, and how the overall measurement system accuracy is calculated from the individual error components. Ways of presenting this data in a useful format for future study and use will now be discussed. The alternative presentations available are tabular and graphical formats, each of which has certain merits compared with the other. In many data collection exercises, part of the measurements and calculations are expressed in tabular form and part graphically, making best use of the merits of each technique.

Tabular data presentation

A tabular presentation allows data values to be recorded in a precise way which maintains exactly the accuracy to which the data values were measured. In other words, the data values are written down exactly as measured. Besides recording the raw data values as measured, tables often also contain further values calculated from the raw data.

An example of a tabular data presentation is given in Table 7.1. This records the results of an experiment to determine the strain induced in a bar of material under a range of stresses. Data were obtained by applying a sequence of forces to the end of the bar and using an extensometer to measure the change in length. Values of the stress and strain in the bar are calculated from these measurements and are also included in the table. The final row, which is of crucial importance in any tabular presentation, is the estimate of possible error in each calculated result.

Table 7.1 Sample tabular presentation of data

Table of measured applied forces and extensometer readings and calculations of stress and strain

Force applied (kN)	Extensometer reading (divisions)	Stress (N/m^2)	Strain
0	0	0	0
2	4.0	15.5	19.8×10^{-5}
4	5.8	31.0	28.6×10^{-5}
6	7.4	46.5	36.6×10^{-5}
8	9.0	62.0	44.4×10^{-5}
10	10.6	77.5	52.4×10^{-5}
12	12.2	93.0	60.2×10^{-5}
14	13.7	108.5	67.6×10^{-5}
Possible error in measurements (%)	± 0.2	± 0.2	± 1.5	$\pm 1.0 \times 10^{-5}$

A table of measurements and calculations should conform to several rules as illustrated in Table 7.1:

1. The table should have a title which explains what data are being presented within the table.

2. Each column of figures in the table should refer to the measurements or calculations associated with one quantity only.
3. Each column of figures should be headed by a title which identifies the data values contained in the column.
4. The units in which quantities in each column are measured should be stated at the top of the column.
5. All titles and columns should be separated by bold horizontal (and sometimes vertical) lines.
6. The errors associated with each data value quoted in the table should be given. The form shown in Table 7.1 is a suitable way to do this when the error level is the same for all data values in a particular column. If error levels vary, however, then it is preferable to write the error boundaries alongside each entry in the table.

Graphical presentation of data

Presentation of data in graphic form involves some compromise in the accuracy to which the data are recorded, as the exact values of measurements are lost. However, graphic presentation has the following two important advantages over tabular presentation:

1. Graphs provide a pictorial representation of results which is more readily comprehended than a set of tabular results.
2. Graphs are particularly useful for expressing the quantitative significance of results and showing whether a linear relationship exists between two variables.

Figure 7.3 shows a graph drawn from the stress and strain values given in Table 7.1. Construction of the graph involves first marking the points corresponding to the stress and strain values. The next step is to draw some line through these data points which best represents the relationship between the two variables. This line will normally be either a straight one or a smooth curve. The data points will not usually lie exactly on this line but, instead, will lie on either side of it. The magnitude of the excursions of the data points from the line drawn will depend on the magnitude of the random measurement errors associated with the data.

As for tables, certain rules exist for the proper representation of data in graphic form:

1. The graph should have a title or caption which explains what data are being presented in the graph.

2. Both axes of the graph should be labelled to express clearly what variable is associated with each axis and to define the units in which the variables are expressed.
3. The number of points marked along each axis should be kept reasonably small – about five divisions is often a suitable number.
4. No attempt should be made to draw the graph outside the boundaries corresponding to the maximum and minimum data values measured, i.e. in Figure 7.3, the graph stops at a point corresponding to the highest measured stress value of 108.5.

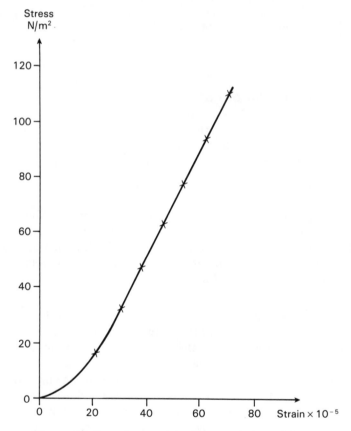

Figure 7.3 Sample graphical presentation of data

Fitting curves to data points on a graph

The procedure of drawing a straight line or smooth curve as appropriate which passes close to all data points on a graph, rather than joining the data points by a jagged line which passes through each data point, is justified on account of the random errors which are known to affect measurements. Any line between the data points is mathematically acceptable as a graphical representation of the data if the maximum deviation of any data point from the line is within the boundaries of the identified level of possible measurement errors. However, within the range of possible lines which could be drawn, only one will be optimum. This optimum line is where the sum of negative errors in data points on one side of the line is balanced by the sum of positive errors in data points on the other side of the line. The nature of the data points is often such that a perfectly acceptable approximation to the optimum can be obtained by drawing a line through the data points by eye. In other cases, however, it is necessary to fit a line by regression techniques (see Appendix 6 for further details of regression techniques).

8

Reliability

8.1 Introduction

The earlier chapters in this book have been concerned with the measurement and error-assessment techniques relevant to quality control procedures during manufacture. However, assurance of the quality of a product immediately after it has been delivered is only one of the parameters in the customer-satisfaction equation. Of at least equal importance is the reliability and longevity of the product after manufacture. The benefits accruing from efforts put into quality control procedures during product manufacture will be quickly lost if the product later gains a reputation of unreliability over its subsequent expected working life. Reliability is an essential component in quality assurance and this chapter is therefore concerned with considering the various means available for measuring and maintaining product reliability. A more comprehensive treatment of the subject can be found in reference 1.

Reliability is formally defined as: *the ability of a product to perform its required function within the bounds of specified working conditions for a stated period of time.* Factors such as manufacturing tolerances, quality variations in raw materials used and differing operating conditions all conspire to make the faultless operating life of a product impossible to predict. Such factors are subject to random variation and chance, and therefore the reliabilty of a product cannot be defined in absolute terms. The nearest one can get to an absolute quantification of reliability are quasi-absolute terms such as the *mean-time-between-failures*, which expresses the average time a product works for without failure. Otherwise, reliability has to be expressed as a statistical parameter which defines the probability that no faults will develop in a product over a specified interval of time.

One immediate difficulty which arises when attempts are made to quantify the reliability of a product is defining what should be counted as a failure.

Failures can generally be divided into three categories:

1. *Critical failures* – where failure causes total loss of function in the product.
2. *Major failures* – where failure causes a major loss of function in a product but it can still continue to be used to some extent.
3. *Minor failures* – where failure leaves the product still able to be used to perform its major purpose but with the loss of some convenience function.

In a product such as a transistor radio, failure of the speaker would be regarded as a critical failure because that would render the radio totally useless. Failure of a noise suppression circuit within it would probably be counted as a major failure, as it would make the radio difficult to listen to though some programs might be heard if the user listened carefully. A minor failure might be failure of the tone adjustment control, which would impair the quality of sound reproduction by a small amount but programs could still be heard reasonably well.

It is probably becoming apparent that classification of failures into these three catagories is not necessarily straightforward. Many types of failure lie on the borderline between categories and it is a matter of personal judgement which side of the border they are placed. Nevertheless, deciding what should be counted as a serious failure is a prerequisite in reliability analysis. Once this has been decided, reliability can be quantified either in quasi-absolute or probabalistic terms.

8.2 Reliability quantification in quasi-absolute terms

While reliability is essentially probabilistic in nature, it can be quantified in quasi-absolute terms by the mean-time-between-failures and the mean-time-to-failure parameters. It must be emphasized that these two quantities are only average values calculated over a number of identical examples of a product. The actual values for any particular sample of a product may vary substantially from these mean values.

The *mean time between failures (MTBF)* is a parameter which expresses the mean number of failures which occur in a product over a given period of time. For example, suppose that the history of a robot manipulator is logged over a one-year (365-day) period and the time intervals in days between faults occurring which require repair are as follows:

11 23 27 16 19 32 6 24 13 21 26 15 14 33 29 12 17 22

The mean interval is twenty days which is therefore the mean-time-between-failures. A simpler way to calculate the MTBF is to merely record the total number of failures which occur over a given time interval. For N failures over a time interval t, the failure rate Θ is given by

$$\Theta = \frac{N}{t}$$

The MTBF is the reciprocal of Θ, i.e.

$$MTBF = \frac{1}{\Theta}$$

The *mean time to failure (MTTF)* is a parameter which is associated with products which are discarded at the first failure because it is either impossible or uneconomic to repair them. It expresses the average time before failure occurs, calculated over a number of identical products. Consumer goods such as cheap transistor radios are examples of products which cannot be repaired economically and are therefore discarded at the first major or critical failure. For example, suppose that a batch of twenty radios are put through an accelerated-wear test (in which a long period of typical use is simulated over a much shorter period) and the simulated life in years before serious failure for each radio is as follows:

7 9 13 6 10 11 8 9 14 8 8 12 9 15 11 9 10 12 8 11

The mean of these twenty numbers is 10. Therefore the simulated mean-time-to-failure is ten years.

For many products, MTBF and MTTF figures are both relevant. The nature of a product often means that minor repairable faults will occur at various points in time during its use, and then, after a certain, greater length of time, a catastrophic failure will occur. Over its working lifetime, the frequency of faults is quantified by the MTBF parameter, and the estimated working life before irredeemable failure occurs is quantified by the MTTF parameter.

A further reliability-associated term of importance is the mean-time-to-repair.

The *mean time to repair (MTTR)* expresses the average time needed for repair of a product, calculated over a number of typical faults which are likely to occur in it. Returning to the example of the robot manipulator, suppose that the time taken in hours to repair each of eighteen faults was as follows:

4 1 3 2 1 9 2 1 7 2 3 4 1 3 2 4 4 1

The mean of these values is 3; therefore the mean-time-to-repair was 3 hours.

The relative importance of the mean-time-to-repair parameter varies according to the product to which it is applied. When applied to a critical element in a production process, the mean-time-to-repair is of equal importance to the mean-time-between-failures. What really matters in this case is the proportion of the total available production time which is lost while the critical element is inoperative. Clearly, repair time is of equal importance to the frequency of fault occurrence. Often, an element whose MTBF is low but whose MTTR is also low will be preferable to an alternative element where the MTBF is a little higher but the MTTR is a lot higher.

This argument about the relative importance of the MTBF and MTTR parameters assumes, however, that elements have to be repaired immediately on failure. Where critical elements in production processes do not have a great cost, it is normal practice to keep spare elements on standby to replace elements on the production process as they fail. In this case, the MTTR is of very little importance. The amount of lost production depends on how long it takes to remove and replace a failed element. In this mode of working, the MTBF parameter assumes very much greater importance because the total production time lost in a given interval of time is clearly proportional to the number of faults occurring over the interval. The MTTR would only become important if it became very large and interrupted the supply of standby elements.

The MTBF and MTTR parameters are often expressed in terms of a combined quantity known as the *availability* figure. This measures the proportion of the total time that a product is working, i.e. the proportion of the total time that it is in an unfailed state.

The *availability* of a product is defined as the following ratio:

$$Availability = \frac{Uptime}{Total\ time} = \frac{MTBF}{MTBF + MTTR}$$

Alternatively, the availability can be expressed as

$$Availability = \frac{1}{1 + \Theta \cdot MTTR}$$

where Θ is the failure rate.

In terms of customer satisfaction, the aim must always be to maximize the MTBF figures and minimize the MTTR figure, thus maximizing the availability.

As far as the MTBF (and MTTF figures) are concerned, good design and high quality control standards during manufacture are the appropriate means of maximizing these figures. Good design procedures which mean that faults are easy to repair are also an important factor in reducing the MTTR figure. However, many factors affecting the MTTR are to a large extent outside the manufacturer's control because they are strongly influenced by customer practices. If a customer chooses to do his own repairs, the time taken to effect each repair is governed by the skill of the personnel he employs to do the work and the stocks of spare parts he keeps. If, on the other hand, repair work is entrusted to the manufacturer, then he can do much to reduce the MTTR by ensuring that his maintenance staff are well trained and motivated and respond quickly to breakdown calls. The manufacturer must also maintain an adequate stock of spare parts and have a means of ensuring that parts are delivered speedily as soon as they are requested. The existence of an efficient parts-delivery service is an important contributor to reducing MTTR figures, even where the customer does his own maintenance.

8.3 Failure patterns

The pattern of failure in a product may increase, stay the same or decrease over its life. Material fatigue is a typical reason for the increase in failure rate over the life of a product. In the early part of their lives, when all components are relatively new, many products exhibit a very low incidence of faults. Then, at a later stage, when fatigue and other ageing processes start to have a significant effect, the rate of faults increases and continues to increase thereafter.

Electronic components are typical of products whose rate of fault incidence decreases over a period of time. Most manufacturing defects, such as sub-standard components, poor sealing against contamination, faulty assembly or bad connections, show up very early on in the life of such products. Thereafter, the rate of fault incidence remains at a low and approximately constant level for a substantial period of time. Where such a failure pattern is known to exist, it is normal practice to 'burn in' components until all the components which are likely to fail in this early period of their life have failed. This should mean that by the time a product is delivered to the customer, it has reached a stage where the rate of fault incidence is constant. If this screening process is controlled correctly, customers will never experience the failure pattern where the rate of fault incidence decreases with time.

A constant pattern of failure over the whole lifetime of the product is typically exhibited by complex systems containing many different components. The various components within such systems each have their own failure pattern where the failure rate is increasing or decreasing with time. The greater the number of such components within a system, the greater is the tendency for the failure patterns in the individual components to cancel out and the rate of fault incidence to assume a constant value.

It is very important to stress at this stage that the MTBF, MTTF and MTTR figures discussed are average values. The greater the number of products over which the average is calculated, the greater will be the accuracy of the figure derived. Any particular sample of a product, however, may have MTBF, etc., figures which differ significantly from the mean value.

Many types of product exhibit all three patterns of failure listed above over some period of their working lives, and have a failure rate–age curve of the form shown in Figure 8.1(a). This is frequently referred to as the 'bathtub curve'. Manufacturers would normally burn in such products before delivery to the customer, to move the failure pattern beyond the left-hand shaded region in Figure 8.1(a) in which the failure rate is decreasing from a high initial value.

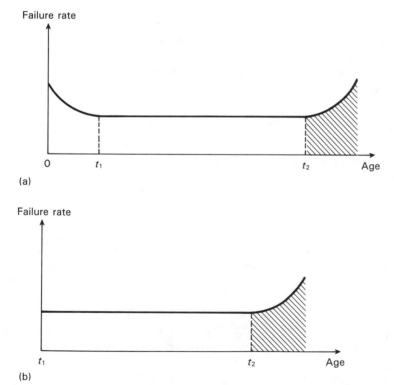

Figure 8.1 'Bathtub curves'

Figure 8.1: (a) typical failure rate–age characteristic for product after manufacture; (b) target failure rate–age characteristic for product after delivery to customer

The failure pattern of the product after delivery, as perceived by the customer, would therefore be as shown in Figure 8.1(b). Normal practice would be to replace the product either when its reliability reaches the right-hand shaded region in Figure 8.1 or shortly afterwards. This ensures that products are replaced before the rate of fault-incidence reaches a high level.

8.4 Reliability quantification in probabilistic terms

It has already been explained how reliability can be expressed in quasi-absolute terms as the mean-time-between-failures. If the average number of failures in a given time for a certain product is Θ, the MTBF can be expressed as

$$MTBF = \frac{1}{\Theta} \tag{8.1}$$

In probabilistic terms, the reliability R_x of a product x is defined as the probability that the product will not fail within a certain period of time. The unreliability U_x is a corresponding term which expresses the probability that the product will fail within a certain time interval. U_x and R_x are related by the expression

$$U_x = 1 - R_x \tag{8.2}$$

For a given time interval t, U_x is related to Θ (and hence to the MTBF) by the expression

$$U_x = 1 - \exp(-\Theta t) \tag{8.3}$$

(It must be noted that this expression is only valid if the failure pattern is in the centre region of the bathtub curve, i.e. the failure rate is a constant. Also, the product must be in a working state at time $t = 0$, i.e. it must be delivered without faults.)

Examination of equation (8.3) shows that at time $t = 0$ the unreliability is zero. Also, as $t \to \infty$, the unreliability tends to a value of 1. This agrees with intuitive expectations that the value of unreliability should lie between values of 0 and 1.

Another point of interest in equation (8.3), is to consider the unreliability when $t = MTBF$, i.e. $t = 1/\Theta$. Then

$$U_x = 1 - \exp(-1) = 0.63$$

i.e. the probability of a product failing after it has been operating for a length of time equal to the MTBF is 63 per cent.

Further analysis of equation (8.3) shows that, for $\Theta t \leqslant 0.1$,

$$U_x \simeq \Theta t \quad ; \quad [\Theta t \leqslant 0.1] \tag{8.4}$$

This is a useful formula for calculating (approximately) the reliability of a critical product which is only used for a time which is a small proportion of its MTBF.

8.5 Laws of reliability in complex systems

If a piece of equipment contains 100 integrated circuits which each have a mean failure rate of one failure per 10^6 h, will the equipment operate for 1000 h without failure? This is clearly a question which a customer would like to see answered, and it is equally clear that it is impossible to give an answer in absolute terms. All that can be done is to express the likelihood or probability that the equipment will work for 1000 h without failure. However, before doing that, it is necessary to introduce some further rules about probability beyond the elementary concepts discussed in Section 7.1.

Probability of the complement [$P(\bar{A})$]

This is the probabilty of an event A not occurring and is given by

$$P(\bar{A}) = 1 - P(A) \tag{8.5}$$

where $P(A)$ is the probability of event A occurring, expressed as a number in the range from 0.0 to 1.0 or as a percentage in the range from 0 per cent to 100 per cent. If there is a 5 per cent chance that a light bulb will fail in the first 500 h of operation [$P(A) = 0.05$], then from equation (8.5), $P(\bar{A}) = 0.95$, i.e. there is a 95 per cent probability that the light bulb will not fail in the first 500 h of operation.

Joint probability [$P(AB)$]

The joint probability is the probability of two independent events A and B both occurring together and is given by

$$P(AB) = P(A) \cdot P(B) \tag{8.6}$$

where $P(A)$ and $P(B)$ are the probabilities, respectively, of the events A and B. Thus the joint probability is the product of the individual probabilities. Expression (8.6) is known as the product rule or series rule and can be extended to cover any number of independent events, i.e.

$$P(AB \cdot \ldots \cdot N) = P(A) \cdot P(B) \cdot \ldots \cdot P(N) \tag{8.7}$$

Single event probability $[P(A + B)]$

This is the probability of any one of two independent events A and B (but not both) occurring, and is given by

$$P(A + B) = P(A) + P(B) - P(AB) \tag{8.8}$$

If A and B are mutually exclusive, i.e. they cannot both occur together, then this expression simplifies to

$$P(A + B) = P(A) + P(B) \tag{8.9}$$

Conditional probability $[P(A \mid B)]$

This is the probability of event A occurring, given that event B has occurred; it is expressed by the notation $P(A \mid B)$. The conditional probability of an event A can be evaluated from

$$P(A \mid B) = \frac{P(AB)}{P(B)} \tag{8.10}$$

In general, $P(A \mid B) < P(A)$. If

$$P(A \mid B) = P(A \mid \bar{B}) = P(A)$$

this shows that $P(A)$ is unaffected by whether or not B occurs, i.e. A and B are independent events.

These principles can be extended to express the reliability of a system containing multiple components which are either in series or in parallel.

Reliability of components in series

In many systems containing multiple components, the whole system fails if any one component within it fails. A good example of this is a system of fairy lights connected in series on a Christmas tree. One way in which the reliability of such a system of series components can be quantified is in terms of the probability that none of the components will fail within a given interval of time.

Using equation (8.7) the reliability R_s of a system of n series components can be expressed as the product of the separate reliabilities of the individual components

$$R_s = R_1R_2R_3 \ldots R_n \tag{8.11}$$

In the case of n identical system components, equation (8.11) simplifies to

$$R_s = (R_x)^n \tag{8.12}$$

where R_x is the reliability of each component.

In the case of the integrated circuits mentioned at the start of Section 8.5, the probability of any individual chip failing within 1000 h of operation is 0.1 per cent. The reliability of each component is, therefore (from equation (8.2)),

$$1.0 - 0.001 = 0.999$$

As all the transistors in the system are nominally identical, equation (8.12) is applicable and the system reliability can be expressed as

$$R_s = (0.999)^{100} = 0.905$$

Thus, there is a 90.5 per cent probability that the system will operate for 1000 h without failure.

Reliability of components in parallel

In many systems containing components connected in parallel, total system failure only occurs when all the elements fail. Street lights are an example of such a system where all lamps are connected in parallel onto the mains electrical supply. The street is only totally dark if all the lamps fail.

For such systems, the system reliability, R_s is given by

$$R_s = 1 - U_s \tag{8.13}$$

where U_s is the unreliability of the system. U_s is calculated in a similar manner to equations (8.11) and (8.12):

$$U_s = U_1 U_2 U_3 \dots U_n \tag{8.14}$$

or, for identical system components,

$$U_s = (U_x)^n \tag{8.15}$$

From equation (8.15),

$$R_s = (1 - U_s) = 1 - U_x^n = 1 - (1 - R_x)^n \tag{8.16}$$

As an example of the use of these equations, consider a lighting system containing four lamps connected in parallel. What is the probability of total system failure within the first 1000 h of operation, given that the reliability of a single lamp over 1000 h is 90 per cent? All components in the system are identical, and therefore the system reliability can be calculated from equations (8.13) and (8.15):

$$U_x = 1 - R_x = 0.1$$

Hence,

$$U_s = (0.1)^4 = 0.0001$$

Thus, the probability of total system failure (no light at all) over 1000 h of operation is 0.01 per cent or, expressed otherwise, the system reliability is 99.99 per cent.

Binomial law of reliability

This law is applicable to systems of parallel components where some component failure can be tolerated, but system failure occurs before all individual components within it have failed.

If R and U are the reliability and unreliability of n identical components connected in parallel in a system, then, by the binomial law of reliability,

$$(R + U)^n = 1 \tag{8.17}$$

Expanding equation (8.17):

$$R^n + nR^{n-1}U + \frac{n(n-1)}{1 \cdot 2} R^{n-2}U^2 + \ldots + \frac{n(n-1)(n-2)\ldots(n-r)}{1 \cdot 2 \cdot 3 \cdot \ldots \cdot r} R^{n-r}U^r + \ldots + U^n = 1$$

(8.18)

If system does not fail unless all n components fail, then

$$R_s = 1 - U^n$$

(8.19)

If system fails when r components fail, then

$$R_s = R^n + nR^{n-1}U + \ldots + \frac{n(n-1)(n-2)\ldots(n-r-1)}{1 \cdot 2 \cdot 3 \cdot \ldots \cdot (r-1)} U^{n-1}$$

(8.20)

and since $R_s = 1 - U_s$,

$$U_s = \frac{n(n-1)(n-2)\ldots(n-r)}{1 \cdot 2 \cdot 3 \cdot \ldots \cdot r} R^{n-r}U^r + \ldots + U^n$$

(8.21)

8.6 Availability of complex systems

The behaviour of complex systems containing many components are frequently described in terms of their availability rather than their reliability. This is because the availability figure is more directly relevant to production cost calculations. The equations to calculate the availabilities of systems containing series or parallel components have an identical form to those for calculating the corresponding reliabilities. Thus, for two series components:

$$A_S = A_1 \times A_2$$

and, for two parallel components:

$$A_S = A_1 + A_2 - (A_1 \times A_2)$$

where A_S is the net system reliability and A_1 and A_2 are the reliabilities of the separate system components.

8.7 Reliability calculations for sample systems

Example 1
The environment in a computer room is controlled by three identical air-conditioning units connected in parallel, as shown in Figure 8.2. The computer can continue to function if only one of the three units is working. If the reliability of each air-conditioning unit is 90 per cent over 10 000 h of operation, what is the overall reliability of the system?

For each unit, $R = 0.9$ and $U = 0.1$

Applying equation (8.20), with $n = 3$ and $r = 1$:

$$R_s = (0.9)^3 + 3(0.9)^2(0.1) + 3(0.9)(0.1)^2 = 0.999$$

Thus, the overall system reliability is 99.9 per cent, i.e the probability that all three air-conditioning units will break down and leave the computer unable to run is only 0.01 per cent.

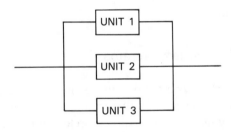

Figure 8.2 Identical units in parallel

Example 2
In a machine shop, finished goods can be produced by either of two parallel routes A and B as shown in Figure 8.3. If the reliabilities of the machines are R_1, R_2, R_3, R_4, what is the total system reliability? (The system is only regarded as failed if neither of the two routes A and B is operational.) From equation (8.11):

$$R_A = R_1 R_2$$

$$R_B = R_3 R_4$$

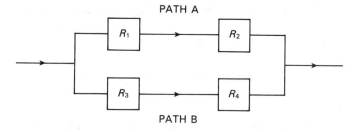

Figure 8.3 Parallel production routes

From equation (8.2):

$$U_A = 1 - R_A = (1 - R_1R_2)$$

$$U_B = 1 - R_B = (1 - R_3R_4)$$

From equation (8.13):

$$R_s = 1 - U_s$$

From equation (8.14):

$$U_s = U_A U_B = (1 - R_1R_2)(1 - R_3R_4)$$

Hence,

$$R_s = 1 - (1 - R_1R_2)(1 - R_3R_4)$$

Example 3
Figure 8.4 shows a block diagram of some units in a production system which are connected in series. The reliabilities of the units for 1000 h of operation are as follows:

$$R_1 = R_3 = R_5 = 0.99 \quad ; \quad R_2 = R_4 = 0.90$$

Figure 8.4 System of production units in series

What is the overall system reliability? Applying equation (8.11):

$$R_s = (0.99)^3 (0.90)^2 = 0.786$$

Thus, the overall system reliability is 78.6 per cent.

Example 4
Figure 8.5(a) shows a block diagram of the same production system as
in Example 3 but where the least reliable units in the system are
duplicated. A switching system automatically transfers production to
the standby unit when one of the duplicated units fails. If the values of
R for each unit are the same as in Example 3, what is the new system
reliability?

The system is equivalent to the system shown in Figure 8.5(b),
where R_A and R_B represent the net reliabilities of the pairs of identical
production units.

From equation (8.16):

$$R_A = R_B = 1 - (1 - 0.90)^2 = 1 - 0.01 = 0.99$$

Thus,

$$R_s = R_1 R_A R_3 R_B R_5 = (0.99)^5 = 0.951$$

Therefore, the overall system reliability is now 95.1 per cent.

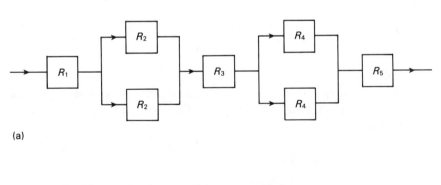

(a)

(b)

Figure 8.5: (a) System of production units in series with duplicated units; (b) Alternative diagram for calculating net efficiency of system (a) above

8.8 Achieving high reliability

Reliability considerations must be at the forefront of the designer's mind whenever he/she is developing a new product or modifying an existing one. The design of the product should be such that manufacturing defects are minimized and the incidence of faults during subsequent use of the product is kept to as low a level as possible. However, there is always an over-riding constraint that, while the finished product must perform the task required of it reliably, the cost of producing it must be sensible with respect to the prevailing market conditions. Thus, features designed to enhance reliability can only be included in a product if their cost does not make the product too expensive for the market it has to compete in. The arguments put forward in Chapter 2 regarding specification of the most cost-effective level of quality control apply equally well to reliability considerations.

Reliability and quality control are strongly correlated. Good quality control during the manufacture of a product is of fundamental importance in achieving high reliability when the product is put into use. The more tightly the production parameters in the manufacture of a product can be constrained to their target values, the fewer weaknesses will be inherent in the product and the less likely it will be to fail. Additionally, if the product being manufactured is identified as having a failure pattern which initially decreases with time, a burn-in procedure should be followed so that its failure pattern has reached steady-state before it is delivered to the customer.

Sound analysis at the design stage, careful quality control during manufacture and burn-in periods after manufacture where appropriate will ensure that the reliability of any particular product is the best that can practically be attained. In their subsequent use, however, many products form part of a larger system, and the requirement then is to maximize the reliability of the whole system at minimum cost.

Several general rules about enhancing the reliability of a product containing several components become apparent if the examples in the last section (8.7) are studied. Firstly, it follows from Example 3 (and also from equation (8.11)) that the greater the number of series components in a system, the poorer the overall system reliability tends to be. Therefore, attempts must be made at the design stage of such a system to include as few separate series components as possible, and, having achieved this, Example 4 shows that, by identifying the critical components in the system which have the worst reliability and duplicating them with identical components connected in parallel, the system reliability can be improved substantially. If this practice is followed, provision must be provided for replacing failed components by standby units. The most efficient way of doing this is via an automatic switching system, but manual methods of replacement can also often work reasonably well. Similar principles of duplication apply in such areas as designing lighting systems. In a particular room, calculations may show that a single 80 W fluorescent tube mounted centrally will provide adequate lighting. If reliability considerations are brought in, however, the designer will specify two 40 W fluorescent tubes rather than a single 80 W tube because this still provides some lighting even if one tube fails. The gain in system reliability is very much greater than the cost of achieving it because the energy consumption will be the same in either case and the only increased cost is the difference in cost between two 40 W tubes and an 80 W tube. The general principle ensuing from this is that a system containing two or more identical components in parallel is usually more reliable than one which contains only a single component.

The principle of increasing reliability by placing components in parallel is often extended by deliberately putting more components in a system than it needs to function at 100 per cent efficiency. This practice is known by the term *'redundancy'*. It is commonly applied in electrical circuits where bad connections are a frequent cause of malfunction. Where connections are made by plugs and sockets, it is often arranged that the same connection is made by two separate pairs of plugs and sockets. The second pair is redundant, i.e. it is not normally needed and the system can function at 100 per cent efficiency without it, but it becomes useful if the first pair fails. A common example of this is the multiple earth connections to be found in a car, where, in good conditions, only one connection is theoretically necessary.

To summarize, then, reliability considerations should be dominant at the design stage of a product. The design should be such that the product is as easy to manufacture as possible. This will ensure that the incidence of faults created in products during manufacture will be minimized. High standards of quality control are also, of course, a necessary condition for this. Finally, the general rule applies that, in a complex system, reliability is maximized when the number of series components is minimized and the number of parallel components which each perform the same function is as large as possible.

9

Practical implementation of measurement and calibration procedures

Earlier chapters have been concerned with setting up the theoretical framework of the measurement and calibration procedures necessary for operation of a quality control and assurance system. Having laid down these theoretical foundations, it is now possible to go on to examine the practical aspects involved in implementing the measurement and calibration procedures required.

9.1 Choice of primary measuring instruments

The initial step in setting up a measurement and calibration system associated with a quality control system is to specify what physical parameters in the manufacturing process have to be measured. In addition, the accuracy with which each parameter must be measured and the environmental conditions in which the measuring instruments must operate are further vital pieces of information. Provision of this information clearly requires expert knowledge from personnel who are intimately acquainted with the operation of the manufacturing plant in question.

Once this information has been acquired, the next stage is to choose suitable instruments for measuring each of the relevant process parameters in the manufacturing system. This again requires the use of an expert instrumentation engineer who has knowledge of all the instruments available for measuring each physical quantity and is able to evaluate their accuracy, cost and suitability for operation in the environmental conditions pertaining. As far as possible, measurement systems and instruments should be chosen which are as insensitive as possible to the operating environment, although this requirement is often difficult to meet because of cost and other performance considerations. Maintainability and constancy of performance are also important criteria in this evaluation because the instrument chosen should be capable of operating for long periods without performance degradation and a requirement for costly maintenance. In consequence of this, the initial cost of an instrument often has a low weighting in the evaluation exercise. Many books are available which give valuable assistance in this evaluation by providing lists and data about all the instruments available for measuring a range of physical quantities (e.g. reference 1 and later chapters in this text). New techniques and instruments are being developed all the time, however, and the instrumentation engineer must thus keep abreast of the latest developments by also reading the appropriate technical journals regularly.

9.2 Choice of secondary measuring instruments

Once all the primary measuring instruments have been specified as described above, the instrumentation engineer must look carefully at their operating characteristics and especially at their behaviour in the environmental conditions relevant to the manufacturing operation. Any susceptibility of their perform- ance characteristics to variations in the environmental conditions must be iden- tified and quantified. Appropriate instruments must then be chosen to measure the relevant environmental parameters so that appropriate corrections can be made to the measurements obtained from the primary measuring instruments. Choice of secondary instruments follows similar principles to that for primary instruments, by choosing an instrument with suitable accuracy, characteristics and cost from the range available.

9.3 Calculation of measurement accuracy

Quality control requires that all manufacturing-process-related measurements should conform to specified and documented measurement limits. This means that the accuracy of each individual measurement, and also the overall accuracy of measurement systems consisting of several components, must all be quantified.

The starting point for calculating measurement accuracy in a system is to identify and quantify the various sources of measurement error present, as described in Chapter 5. In situations where there are significant modifying inputs in the system, knowledge of the operational characteristics of the measuring instruments used, as discussed in Chapter 4, is also important. This then allows the readings obtained from secondary transducers to be used to correct the outputs from the primary measuring instruments. Other processing of the output signals from the primary measuring instruments is also carried out, as described in Chapter 6, to remove bias and noise, etc., and ensure that the accuracy of each measurement is the best that is obtainable. Having calculated the separate accuracies of measuring system components in this manner, the overall accuracy of a measurement system containing several components can then be calculated according to the formulae presented in Chapter 7.

When measurement accuracy figures are written down in the quality system documentation, the figures quoted must reflect the maximum errors which may exist in measurements. Such maximum errors will occur when instruments have drifted furthest from their quoted characteristics just prior to re-calibration. Knowledge of the way in which instrument characteristics drift and the calibration frequency set is obviously required before this maximum error can be defined. This is then the magnitude of measurement error which the calibration system guarantees will not be exceeded provided that the calibration procedures and frequency are followed correctly.

To summarize, therefore, the quoted accuracy level for a measurement system is derived by the following steps:

1. Establish the characteristics of all instruments under controlled environmental conditions by calibrating them against standards instruments traceable back to reference standards.
2. Establish the maximum variation in instrument error levels at the extremes of possible operating conditions and just prior to the instrument being calibrated, thus setting the error for each measurement system component.
3. Calculate the error-band for the whole measurement system from error-bands of individual system components.

9.4 Calibration procedures

A fundamental requirement in all quality control and assurance systems is that calibration procedures be implemented to monitor the accuracy of all measurement systems employed in order that the specified quality level be maintained. Such periodic calibration of instruments ensures that their output reading will be within stated measurement-limit bounds when they are used under the environmental conditions specified for the calibration exercise.

Every instrument in the workplace has a designated person who is responsible for it. That person must ensure that all instruments for which he/she is responsible are calibrated at the correct times by approved personnel. Approved personnel means either staff within the company who have attended all the necessary courses relevant to the calibration duties or sub-contractors outside the company who are verified as being able to provide calibration services satisfactorily.

The first step in instrument calibration is to establish all the inputs to the instrument (desired and modifying) which have an effect on the output. Keeping all modifying inputs at some specified level, the desired input is then varied in steps over some range of values, causing the instrument output to vary in steps over a corresponding range of readings. At the same time, the varying input is also applied to a 'standard' instrument whose output is compared with the values registered by the instrument being calibrated. The 'standard' instrument referred to is a high-quality one kept solely for calibration duties, whose accuracy is at least ten times better than the nominal value of the instrument being calibrated. The 'standard' instrument must form part of a calibration chain and be calibrated itself against a still more accurate standard. A typical calibration chain was shown in Figure 3.1. Adherence to this practice fulfils the requirement that the accuracy of all process instruments and the 'standard' instruments used to calibrate them be *traceable* back to the fundamental standards maintained by the National Physical Laboratory (in the United Kingdom), or other appropriate national body responsible for the maintenance of measurement standards.

Effect of modifying inputs

It was mentioned above that calibration guarantees that the accuracy of the output reading in a calibrated instrument will be at a certain level when the instrument is used under stated environmental conditions. Outside those conditions, the accuracy of the instrument will vary to a greater or lesser extent according to its susceptibility to the modifying inputs inherent in a varying environment. In all normal measurement situations, it is impossible to control the environmental conditions to be at the levels specified for calibration, and thus due account must be taken of this by appropriate correction of the primary measuring instrument output readings.

The choice of suitable secondary measuring instruments to monitor the level of modifying inputs has already been discussed in Section 9.2. However, apart from knowing the magnitude of modifying inputs, it is also necessary to quantify their effect on the primary measurement. The effect of modifying inputs is determined in a very similar manner to the calibration procedure described above. The level of one modifying input is varied in steps over a certain range while the level of the desired input (primary measurement) and all other modifying inputs are held constant, and the corresponding output readings are recorded. This allows an input–output relationship to be drawn for that particular modifying input. The same procedure is then repeated for all other modifying inputs.

This procedure represents a statement of a desired ideal rather than a realistic picture of what is normally carried out in practice. It is usual to disregard modifying inputs which only have a small effect on the output of measuring instruments, as the measurement errors introduced by so doing are insignificant. Changes in atmospheric pressure are typical of such modifying inputs which are generally disregarded.

Rate of change of instrument characteristics

The discussion so far has been concerned with specifying what the characteristics of an instrument are immediately following calibration. These characteristics will change over a period of time, because of factors such as ageing effects, mechanical wear, long-term environmental changes, and the effects of dust, fumes and chemicals in the operating atmosphere. The extent and rate of this change must be determined so that the appropriate time when the instrument will need re-calibrating can be determined. Instrument manufacturers can give very useful guidance on this, but their figures can only be approximate, as the rate of variation in characteristics will be affected by the operating environment in the particular application of the measurement system. Some practical investigation of the rate of variation of characteristics will therefore always be necessary.

Determining the calibration frequency required

When installing instruments in a new system, or new instruments in an existing system, it is highly likely that much thought will have been given to the suitability of the instruments chosen for their intended application. Knowledge of the instruments' characteristics and their operating environments will normally allow good approximations to be made as to the rate at which their performance will change with time and therefore what frequency of re-calibration is required.

Because good instrument calibration is so essential to the proper functioning of quality control systems, however, it is prudent not to rely too much on such *a priori* predictions. It is always possible that some very significant factor in the procedure for estimating the required calibration frequency has been inadvertently neglected. Therefore, as long as the circumstances permit, it is sensible to start from basics in deriving the required calibration frequency and not use past-history information about the instrument and its operating environment. Applying this philosophy, the following frequency of checking instrument characteristics might be appropriate (assuming 24 h/day working):

Day 1	once per hour for first 4 hours, then every 4 hours
Days 2–3	every 12 hours
Days 4–7	once per day
Weeks 2–3	twice per week
Weeks 3–4	once per week
Week 6	once
Week 8	once
Months 3–6	once per month
Month 9	once
Month 12	once
Month 18	once
Month 24	once
Then once per year thereafter.	

The above reducing frequency of calibration checks should be applied until a point is reached where a deterioration in the instrument's accuracy is first detected. Comparison of the amount of performance degradation with the inaccuracy level that is permissible in the instrument will show whether the instrument should be calibrated immediately at this point or whether it can be safely left for a further period. If the above pattern of calibration checks were followed, and the check at Week 8 showed a deterioration in accuracy which was close to the permissible limit, this would determine that the calibration frequency for the instrument should be every eight weeks.

The above method of establishing the optimum calibration frequency is clearly an ideal which cannot always be achieved in practice, and indeed for some types of instrument this level of rigor is unnecessary. When used on many production processes, for instance, it would be unacceptable to interrupt production every hour to re-check instrument calibrations unless a very good case could be made as to why this was necessary. Also, the nature of some instruments, for example a mercury-in-glass thermometer, means that calibration checks will only ever be required infrequently. However, for instruments such as unprotected base-metal thermocouples, initial calibration checks after one hour of operation would not be at all inappropriate.

9.5 Procedure following calibration

When the instrument is calibrated against a standard instrument, its accuracy will be shown to be either inside or outside the required measurement accuracy limits. If the instrument is found to be inside the required measurement limits, the only course of action required is to record the calibration results in the instrument's record sheet and then put it back into use until the next scheduled time for calibration.

The options available if the instrument is found to be outside the required measurement limits depends on whether its characteristics can be adjusted and the extent to which this is possible. If the instrument has adjustment screws, these should be turned until the characteristics of the instrument are within the specified measurement limits. Following this, the adjustment screws must be sealed to prevent tampering during the instrument's subsequent use. The instrument can then be returned to its normal operating position for further use.

The second possible course of action if the instrument is outside measurement limits covers the case where no adjustment is possible or the range of possible adjustment is insufficient to bring the instrument back within measurement limits. In this event, the instrument must be withdrawn from use, and this withdrawal must be marked prominently on it to prevent inadvertent re-use. The options then available are to either send the instrument for repair if this is feasible or to scrap it.

9.6 Documentation of measurement and calibration systems

All measurement and calibration systems implemented must be fully documented. In the case of a small company, all the relevant information should be contained within one manual, whereas for a large company it is often more appropriate to have separate volumes covering corporate and division procedures respectively.

The manual (or manuals) documenting the measurement and calibration procedures associated with the quality control systems in operation should contain the following information.

Documentation revision level and distribution

1. The date and revision level of the manual.
2. The number of copies of the manual issued and the names of the authorized personnel holding them.
3. The procedure for amendment of the manual.

Measurement systems

4. A full description of the measurement requirements throughout the workplace.
5. The measurement limits required for each measurement system.
6. The system of serial numbers used to uniquely identify all instruments used in the workplace.
7. A list of the instruments used and the name of the person responsible for each.
8. A description of each type of instrument used.
9. The required calibration frequency for each instrument (where for some reason calibration is not required for any particular instrument, the documentation must specifically state that it is excluded).
10. Instructions as to the correct way of using instruments.
11. Information about any environmental control or other special precautions to be taken when using instruments.
12. Training courses (if any) to be attended by persons using instruments.

Calibration procedures

13. The standard instruments to be used for calibration.
14. The required method of storing and handling standard instruments.
15. The standard environmental conditions required for performance of calibration functions (if a common and well-accepted procedure is published elsewhere, this can be referenced rather than including the procedure in full in the calibration document).
16. The standard format required for the recording of calibration results within the instrument manual (including the date of calibration).
17. The standard format required for recording calibration results on the instruments themselves (where appropriate).
18. The procedure to be followed if an instrument is found to be outside the calibration limits.
19. The method of marking instruments withdrawn from use because they are outside calibration limits.
20. The traceability of the calibration system back to national reference standards
21. Training courses to be attended by personnel performing instrument calibration duties and refresher courses (if any) required.

Calibration reviews

22. The review procedure for the continued effectiveness of the calibration system being operated.
23. The results of each such effectiveness review.

The typical structure of the manual documenting this information is given in Appendix 7.

9.7 Instrument records

A separate record for each instrument in the system must be maintained which specifies as a minimum:

(a) its serial number;
(b) the name of the person responsible for calibrating it;
(c) the required calibration frequency;
(d) the date of the last calibration;
(e) the calibration results.

These sheets should be bound together as a site instrument manual.
   A suitable format for recording this information is given in Table 9.1.

Table 9.1 Typical format for instrument record sheets

Type of instrument:	Company serial number:
Manufacturer's part number:	Manufacturer's serial number:
Measurement limit:	Date introduced:
Location:	
Instructions for use:	
Calibration frequency:	Signature of person responsible for calibration:

CALIBRATION RECORD

Calibration date	Calibration results	Calibrated by

10

Temperature calibration

10.1 Review of temperature measuring instruments

Instruments to measure temperature can be divided into six separate classes according to the physical principle on which they operate. These principles are as follows:

(a) thermal expansion;
(b) the thermoelectric effect;
(c) resistance change;
(d) resonant frequency change;
(e) velocity of sound;
(f) radiative heat emission.

Thermal expansion methods make use of the fact that the dimensions of all substances, whether solids, liquids or gases, change with temperature. Instruments operating on this physical principle include the liquid-in-glass thermometer, the bimetallic thermometer and the pressure thermometer.

Thermoelectric effect instruments rely on the physical principle that, when any two different metals are connected together, an e.m.f., which is a function of the temperature, is generated at the junction between the metals. The general form of this relationship is

$$e = a_1 T + a_2 T^2 + a_3 T^3 + \ldots + a_n T^n$$

This is clearly non-linear, which is inconvenient for measurement applications. Fortunately, for certain pairs of materials, the terms involving squared and higher powers of $T (a_2 T^2, a_3 T^3, \text{etc.})$ are approximately zero and the e.m.f.–temperature relationship is approximately linear according to

$$e \simeq a_1 T$$

Wires of such pairs of materials are connected together at one end, and in this form are known as thermocouples. Thermocouples are a very important class of device as they provide the most commonly used method of measuring temperatures in industry.

Varying-resistance devices rely on the physical principle of the variation of resistance with temperature. The instruments working on this principle are known as either resistance thermometers or thermistors according to whether the material used for their construction is a metal or a semiconductor material.

The principle of resonant frequency change with temperature is used in the quartz thermometer. The resonant frequency of materials such as quartz is a function of temperature, which enables temperature changes to be translated into frequency changes.

Another parameter which changes with temperature is the velocity of sound in a gas. This principle is used in acoustic thermometers.

The final physical principle which can be used to measure temperature is radiation emission from a body. All bodies emit electromagnetic radiation as a function of their temperature above absolute zero. Measurement of the radiation emission therefore allows the temperature of the body to be calculated. Instruments using this principle are known as radiation thermometers.

Liquid-in-glass thermometers

The liquid-in-glass thermometer contains a fluid, either mercury or coloured alcohol, within a bulb and capillary tube, as shown in Figure 10.1. As the temperature rises, the fluid expands along the capillary tube and the meniscus level is read against a calibrated scale etched on the tube. Typically, the measurement inaccuracy of such instruments is quoted as ±1 per cent of full scale. However, an inaccuracy of only ±0.15 per cent can be obtained in the best industrial instruments. Industrial versions of the liquid-in-glass thermometer are normally used to measure temperatures in the range between −200 °C and +1000 °C, although instruments are available to special order which can measure temperatures up to 1500 °C.

The major source of measurement error arises from the difficulty of estimating the position of the curved meniscus of the fluid correctly against the scale. In the longer term, additional errors are introduced due to volumetric changes in the glass. Such changes occur because of creep-like processes in the glass, but occur only over a timescale of years. Annual calibration checks are therefore advisable.

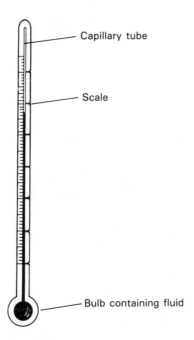

Capillary tube

Scale

Bulb containing fluid

Figure 10.1 The liquid-in-glass thermometer

Bimetallic thermometer

The bimetallic principle is probably more commonly known in connection with its use in thermostats. It is based on the fact that if two strips of different metals are bonded together, any temperature change will cause the strip to bend, as this is the only way in which the differing rates of change of length of each metal in the bonded strip can be accommodated. In the bimetallic thermostat, this is used as a switch in control applications.

If the magnitude of bending is measured, the bimetallic device becomes a thermometer. For such purposes, the strip is often arranged in a spiral con-figuration, as shown in Figure 10.2, as this gives a relatively large displacement of the free end for any given temperature change. Strips in a helical shape are an alternative for this purpose. The measurement sensitivity is increased further by choosing the pair of materials carefully such that the degree of bending is maximized, with Invar (a nickel–steel alloy) and brass being commonly used.

For visual indication purposes, the end of the strip can be made to turn a pointer mounted on low-friction bearings which moves against a calibrated scale. Another suitable form of translational displacement transducer is the LVDT (linear variable differential transformer).

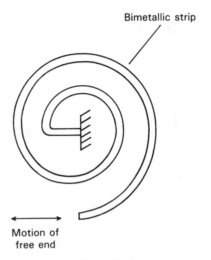

Figure 10.2 Bimetallic thermometer

Bimetallic thermometers are used to measure temperatures between −75 °C and +1500 °C. The inaccuracy of the best instruments can be as little as ±0.5 per cent but such devices are quite expensive. Many instrument applications do not require this degree of accuracy in temperature measurements, and in such cases much cheaper bimetallic thermometers with substantially inferior accuracy specifications are used.

All such devices are liable to suffer changes in characteristics due to contamination of the metal components exposed to the operating environment. Further changes are to be expected arising from mechanical damage during use, particularly if these devices are mishandled or dropped. As the magnitude of these effects varies with their application, the required calibration interval must be determined by practical experimentation.

Pressure thermometers

The pressure thermometer measures the variation in pressure of a gas constrained inside a bulb of fixed volume as the temperature changes. As such, it does not strictly belong to the thermal expansion class of instruments but is included because of the relationship between volume and pressure according to Boyle's gas law.

The change in pressure of the gas is measured by a suitable pressure transducer such as the Bourdon tube. This transducer is located remotely from the bulb and connected to it by a capillary tube, as shown in Figure 10.3. The

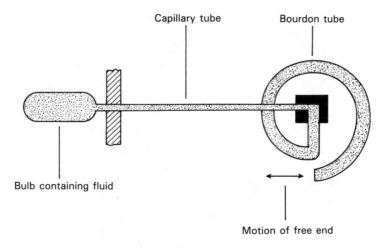

Figure 10.3 The pressure thermometer

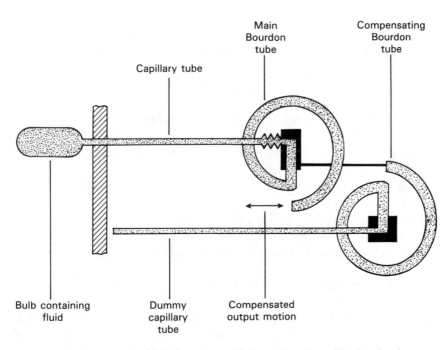

Figure 10.4 Correction for temperature gradient along capillary tube in pressure thermometer

need to protect the pressure-measuring instrument from the environment where the temperature is being measured can require the use of capillary tubes up to 5 m long, and the temperature gradient, and hence pressure gradient, along the tube acts as a modifying input which can introduce a significant measurement error. Correction for this using the principle of introducing an opposing modifying input, as discussed in Chapter 5, can be carried out according to the scheme shown in Figure 10.4. This includes a second, dummy capillary tube whose temperature gradient is measured by a second Bourdon tube. The outputs of the two Bourdon tubes are connected together in such a manner that the output from the second tube is subtracted from the output from the first, thus eliminating the error due to the temperature gradient along the tube.

Pressure thermometers are used to measure temperatures in the range between $-250\,°C$ and $+2000\,°C$ and their typical inaccuracy is ± 0.5 per cent of full-scale reading. Errors occur in the short term due to mechanical damage and in the longer term due to small volumetric changes in the glass components. The rate of increase in these errors is mainly use-related and therefore the required calibration interval must be determined by practical experimentation.

Thermocouples

Thermocouples are manufactured from various combinations of the base metals copper and iron, the base-metal alloys of alumel (Ni–Mn–Al–Si), chromel (Ni–Cr), constantan (Cu–Ni), nicrosil (Ni–Cr–Si) and nisil (Ni–Si–Mn), the noble metals platinum and tungsten, and the noble-metal alloys of platinum–rhodium and tungsten–rhenium.[1] The particular combinations which are used as thermocouples are known by internationally recognized letters, for instance type K being chromel–alumel. The e.m.f.–temperature characteristics for some of these standard thermocouples are shown in Figure 10.5. Certain of these exhibit reasonable linearity over short temperature ranges and their characteristic can therefore be approximated by a series of straight-line relationships for use in intelligent instruments containing thermocouples. In general, however, the relationships do not approximate sufficiently to straight line ones, and the temperature indicated by a given e.m.f. output measurement has to be calculated from published thermocouple tables.

Thermocouple tables are calculated assuming that the reference junction of the thermocouple is at $0\,°C$. This condition can be achieved by immersing the reference junction in an ice bath. However, it is not easy in many applications to maintain such a low reference temperature. In such circumstances, the reference junction can be placed in an environment maintained at some higher temperature by an electrical heating element. Correction then has to be made

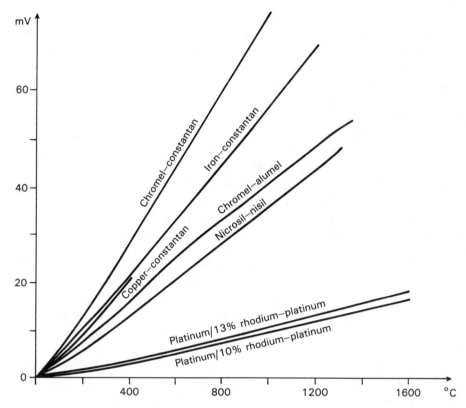

Figure 10.5 E.m.f. – temperature characteristics for some standard thermocouple materials

to the temperature indicated by the thermocouple tables to allow for the different reference junction temperature. The procedure for this is explained elsewhere.[2]

In order to make a thermocouple conform to some precisely defined e.m.f.–temperature characteristic described by standard tables, all metals used must be refined to a high degree of pureness and all alloys manufactured to an exact specification. This makes the materials used expensive, and consequently thermocouples are typically only a few centimetres long. It is clearly impractical to connect a voltage-measuring instrument to measure the thermocouple output in such close proximity to the environment whose temperature is being measured, and therefore extension leads up to several metres long are normally connected between the thermocouple and the measuring instrument. It is important that the voltage measured at the open ends of the extension wires arises only from the hot junction of the thermocouple immersed in the medium whose temperature is being measured. Therefore, no e.m.f. must be generated at the junction between the thermocouple and its extension leads and, consequently, the thermocouple and the extension wires must have very similar thermoelectric properties. This condition is usually met by choosing extension leads of the same basic materials as the thermocouple but manufactured to a lower specification. However, this solution is still prohibitively expensive in the case of noble-metal thermocouples, and base-metal extension leads which have a similar thermoelectric behaviour to the noble-metal thermocouple are generally used. In this form, the extension leads are known as compensating leads. A typical example is the use of nickel/copper–copper extension leads connected to a platinum/rhodium–platinum thermocouple.

The five standard base metal thermocouples are chromel–constantan (type E), iron–constantan (type J), chromel–alumel (type K), nicrosil–nisil (type N) and copper–constantan (type T). These are all relatively cheap to manufacture but they become inaccurate with age and have a short life.

Chromel–constantan devices give the highest measurement sensitivity of $80\,\mu V/^{\circ}C$, with an inaccuracy of ± 0.75 per cent and a useful measuring range of $-200\,^{\circ}C$ up to $900\,^{\circ}C$. Unfortunately, while they can operate satisfactorily in oxidizing environments, their performance and lives are seriously affected by reducing atmospheres. Iron–constantan thermocouples have a sensitivity of $60\,\mu V/^{\circ}C$ and are the preferred type for general purpose measurements in the temperature range $-150\,^{\circ}C$ to $+1000\,^{\circ}C$, where the typical measurement inaccuracy is ± 1 per cent. Their performance is little affected by either oxidizing or reducing atmospheres. Copper–constantan devices have a similar measurement sensitivity of $60\,\mu V/^{\circ}C$ and find their main application in measuring sub-zero temperatures down to $-200\,^{\circ}C$, with an inaccuracy of ± 0.5 per cent. They can also be used in both oxidizing and reducing atmospheres to measure temperatures up to $350\,^{\circ}C$. Chromel–alumel thermocouples have a measurement sensitivity of only $45\,\mu V/^{\circ}C$, although their characteristic is particularly linear over the temperature range between $700\,^{\circ}C$ and $1200\,^{\circ}C$ and this is therefore their main application. Like chromel–constantan devices, they are suitable for oxidizing atmospheres but not for reducing ones. The measurement inaccuracy for them is ± 0.75 per cent. Nicrosil–nisil thermocouples are a recent development which resulted from attempts to improve the performance and stability of chromel–alumel thermocouples. Their thermoelectric characteristic has a very similar shape to type K devices with equally good linearity over the high-temperature measurement range and a measurement sensitivity of $40\,\mu V/^{\circ}C$. The operating environment limitations are the same as for chromel–alumel devices but their long-term stability and life are at least three times better.

Noble-metal thermocouples are always expensive but enjoy high stability and long life in conditions of high temperature and oxidizing environments.

They are chemically inert except in reducing atmospheres. Thermocouples made from platinum and a platinum–rhodium alloy have a low inaccuracy of ± 0.2 per cent and can measure temperatures up to 1500 °C, but their measurement sensitivity is only 10 μV/°C. Alternative devices made from tungsten and a tungsten–rhenium alloy have a better sensitivity of 20 μV/°C and can measure temperatures up to 2300 °C.

Thermocouples are delicate instruments which must be treated carefully if their specified operating characteristics are to be reproduced. One major source of error is induced strain in the hot junction which reduces the e.m.f. output, and precautions are normally taken to minimize this by mounting the thermocouple horizontally rather than vertically. In some operating environments, no further protection is necessary to obtain satisfactory performance from the instrument. However, thermocouples are prone to contamination by various metals and protection is often necessary to minimize this. Such contamination alters the thermoelectric behaviour of the device, such that its characteristic varies from that published in standard tables. It also becomes brittle and its life is therefore shortened.

Protection takes the form of enclosing the thermocouple in a sheath. Some common sheath materials and their maximum operating temperatures are shown in Table 10.1. While the thermocouple is an instrument with a naturally first order type of step response characteristic, the time constant is usually so small as to be negligible when the thermocouple is used unprotected. When enclosed in a sheath, however, the time constant of the combination of thermocouple and sheath is significant. The size of the thermocouple and hence the

Table 10.1 Common sheath materials for thermocouples

Material	Maximum operating temperature (°C)
Mild steel	900
Nickel–chromium	900
Fused silica	1000
Special steel	1100
Mullite	1700
Recrystallized alumina	1850
Beryllia	2300
Magnesia	2400
Zirconia	2400
Thoria	2600

Note: The maximum operating temperatures quoted assume oxidizing or neutral atmospheres. For operation in reducing atmospheres, the maximum allowable temperature is usually reduced.

diameter required for the sheath has a large effect on the importance of this. The time constant of a thermocouple in a 1 mm diameter sheath is only 0.15 s and this has little practical effect in most measurement situations, whereas a larger sheath of 6 mm diameter gives a time constant of 3.9 s which cannot be ignored so easily.

Thermocouples are manufactured by connecting together two wires of different materials, where each material is produced so as to conform precisely with some defined composition specification. This ensures that its thermoelectric behaviour accurately follows that for which standard thermocouple tables apply. The connection between the two wires is effected by welding, soldering or in some cases just by twisting the wire ends together. Welding is the most common technique used generally, with silver-soldering being reserved for copper–constantan instruments.

The diameter of wire used to construct thermocouples is usually in the range between 0.4 mm and 2 mm. The larger diameters are used where ruggedness and long life are required, although these advantages are gained at the expense of increasing the measurement time constant. In the case of noble-metal thermocouples, the use of large-diameter wire incurs a substantial cost penalty. Some special applications have a requirement for a very fast response time in the measurement of temperature, and in such cases wire diameters as small as 0.1 μm (0.1 microns) can be used.

The mode of construction of thermocouples means that their characteristics can be incorrect even when they are new, due to faults in either the homogeneity of the thermocouple materials or in the construction of the device. Therefore, calibration checks should be carried out on all new thermocouples before they are put into use. Thereafter, the rate of change of thermoelectric characteristics with time is entirely dependent upon the operating environment and the degree of exposure to it. Particularly relevant factors in the environment are the type and concentration of trace metal elements and the temperature (the rate of contamination of thermocouple materials with trace elements of metals is a function of temperature). A suitable calibration frequency can therefore only be defined by practical experimentation, and this must be reviewed whenever the operating environment and conditions of use change.

The thermopile

The thermopile is the name given to a temperature-measuring device which consists of several thermocouples connected together in series, such that all the reference junctions are at the same cold temperature and all the hot junctions are exposed to the temperature being measured, as shown in Figure 10.6. The effect of connecting n thermocouples together in series is to increase the

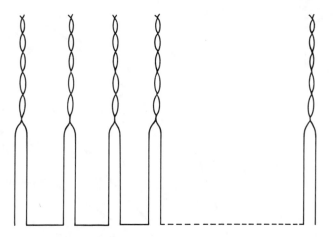

Figure 10.6 The thermopile

measurement sensitivity by a factor of n. A typical thermopile manufactured by connecting together twenty-five chromel–constantan thermocouples gives a measurement resolution of 0.001 $^\circ$C. As the thermopile is essentially a multiple thermocouple, all the previous discussion about thermocouples applies equally well to thermopiles and calibration requirements are also identical.

Resistance thermometers

Resistance thermometers rely on the principle that the resistance of a metal varies with temperature according to the relationship

$$R = R_0(1 + a_1T + a_2T^2 + a_3T^3 + ... + a_nT^n) \tag{10.1}$$

This equation is non-linear and so is inconvenient for measurement purposes. The equation becomes linear if all the terms in a_2T^2 and higher powers of T are negligible. This is approximately true for some metals over a limited temperature range and, in such cases, the resistance and temperature are related according to

$$R \simeq R_0(1 + a_1T) \tag{10.2}$$

Platinum is one such metal where the resistance–temperature relationship is linear within ± 0.4 per cent over the temperature range between $-200\,^\circ$C and $+40\,^\circ$C. Even at $+1000\,^\circ$C, the quoted maximum inaccuracy figure is only

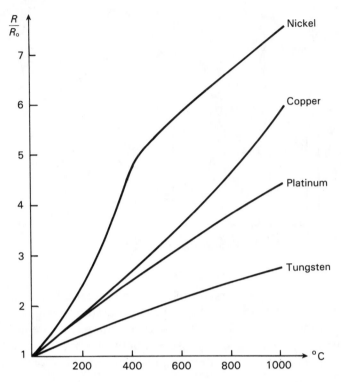

Figure 10.7 Typical resistance–temperature characteristics of metals

±1.2 per cent. The characteristics of platinum and other metals are summarized in Figure 10.7.

Its good linearity and chemical inertness makes platinum the first choice for resistance thermometers in many applications. Platinum is very expensive, however, and consequently the cheaper but less accurate alternatives nickel and copper are sometimes used. These two metals are very susceptible to oxidation and corrosion and so the range of applications for which they can be used is strictly limited even if their reduced accuracy is acceptable. Another metal, tungsten, is also used in some circumstances, particularly for high-temperature measurements. The working range of each of these four types of resistance thermometer are as shown below:

Platinum: −270 °C to 1000 °C (though use above 650 °C is uncommon)
Copper: −200 °C to 260 °C
Nickel: −200 °C to 430 °C
Tungsten: −270 °C to 1100 °C

In the case of non-corrosive and non-conducting environments, resistance thermometers are used without protection. In all other applications, they are protected inside a sheath. As in the case of thermocouples, such protection affects the speed of response of the system to rapid changes in temperature. A typical time constant for a sheathed platinum resistance thermometer is 0.4 s.

The resistance thermometer exists physically as a coil of metal resistance wire. The different instruments available have resistance elements ranging from 10Ω right up to $25\,000\Omega$. The devices with high resistance have several operational advantages. The first of these is that any connection resistances within the circuit become negligible in their effect. The second advantage is that the relatively high-voltage output produced by these instruments makes any induced e.m.f.'s produced by thermoelectric behaviour at the junction with connection leads negligible in magnitude.

The normal method of measuring resistance is to use a dc bridge. The excitation voltage of the bridge has to be chosen very carefully because, although a high value is desirable for achieving high measurement sensitivity, the self-heating effect of high currents flowing in the temperature transducer creates an error by increasing the temperature of the device and so changing the resistance value.

The frequency at which a resistance thermometer should be calibrated depends upon the material it is made from and upon the operating environment. Practical experimentation is therefore needed to determine the necessary frequency and this must be reviewed if the operating conditions change.

Thermistors

Thermistors are manufactured from beads of semiconductor material prepared from oxides of the iron group of metals such as chromium, cobalt, iron, manganese, and nickel. The resistance of such materials varies with temperature according to the following expression:

$$R = R_{0} \cdot \exp[\beta(1/T - 1/T_{0})] \tag{10.3}$$

This relationship, as illustrated in Figure 10.8, exhibits a large negative temperature coefficient (i.e. the resistance decreases as the temperature increases), and so is fundamentally different from the relationship for the resistance thermometer, which shows a positive temperature coefficient. The form of equation (10.3) is such that it is not possible to make a linear approximation to the curve over even a small temperature range, and hence the thermistor is very definitely a non-linear instrument.

The major advantages of thermistors are their relatively low cost and their

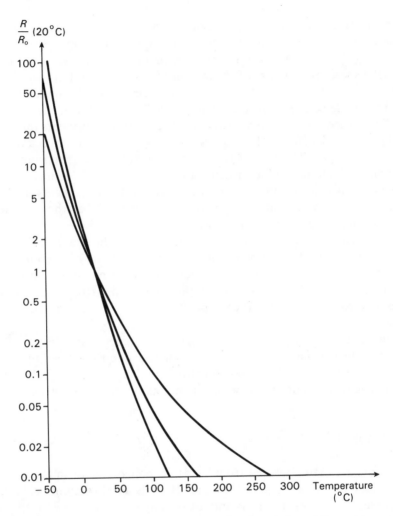

Figure 10.8 Typical resistance–temperature characteristics of thermistor materials

small size. This size advantage means that the time constant of thermistors operated in sheaths is small. However, the size reduction also decreases heat dissipation capability, and so makes the self-heating effect greater. In consequence, thermistors have to be operated at generally lower current levels than resistance thermometers and so the measurement sensitivity afforded is less.

Like the resistance thermometer, the resistance of a thermistor is measured by a dc bridge, and similar comments apply regarding the required calibration frequency.

Quartz thermometers

The quartz thermometer is a fairly expensive device which has been developed recently and makes use of the principle that the resonant frequency of a material such as quartz varies with temperature. The temperature-sensing element consists of a quartz crystal enclosed within a probe (sheath). The crystal is connected electrically so as to form the resonant element within an electronic oscillator. Measurement of the oscillator frequency therefore allows the measured temperature to be calculated.

The instrument has a very linear output characteristic over the temperature range between $-40\,°C$ and $+230\,°C$, with a measurement resolution of $0.1\,°C$. The characteristics of the instrument are generally very stable over long periods of time and therefore only infrequent calibration will be necessary. However, the exact calibration frequency required should be determined experimentally in the usual way.

Acoustic thermometers

The principle of acoustic thermometry was invented as long ago as 1873, but until very recently it has only been used for measuring cryogenic (very low) temperatures. Acoustic thermometers use the fact that the velocity of sound through a gas varies with temperature according to the equation

$$v = (\alpha RT/M)^{1/2} \tag{10.4}$$

where v is the sound velocity, T is the gas temperature, M is the molecular weight of the gas and both R and α are constants. The various versions of acoustic thermometer which are available differ according to the technique used for generating sound and measuring its velocity in the gas. Further information can be found in reference 3.

Radiation thermometers

The term 'radiation thermometer' describes a whole class of devices which measure the radiation emission from a body, and includes the optical pyrometer and various forms of radiation pyrometer. Between them, the various devices cover the temperature range from $-20\,°C$ up to $+1800\,°C$. The total rate of radiation emission per second is given by

$$E = K \cdot T^4 \tag{10.5}$$

where T is the temperature of the body in degrees Kelvin.

The power spectral density of this emission varies with temperature in the manner shown in Figure 10.9. The major part of the frequency spectrum lies within the band of wavelengths between 0.3 μm and 40 μm, which corresponds to the visible (0.3–0.72 μm) and infra-red (0.72–1000 μm) ranges. Choice of the best method of measuring the emitted radiation depends on the temperature of

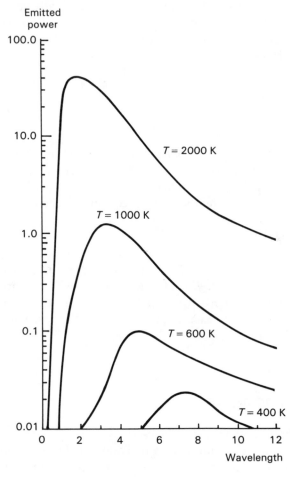

Figure 10.9 Power spectral density of radiated energy emission at various temperatures

the body. At low temperatures, the peak of the power spectral density function lies in the infra-red region, whereas at higher temperatures it moves towards the visible part of the spectrum. This phenomenon is observed as the red glow which a body begins to emit as its temperature is increased beyond 600 °C.

Radiation thermometers have one major advantage in that they do not require to be in contact with the hot body in order to measure its temperature. This makes them especially suitable for measuring high temperatures which are beyond the capabilities of contact instruments such as thermocouples, resistance thermometers and thermistors. They are also capable of measuring moving bodies, for instance the temperature of steel bars in a rolling mill. Their use is not as straightforward as the discussion so far might have suggested, however, because the radiation from a body varies with its composition and surface condition as well as with temperature. This dependence on surface condition is quantified by a parameter known as the *emissivity* of the body. The use of radiation thermometers is further complicated by absorption and scattering of the energy between the emitting body and the radiation detector. Energy is scattered by atmospheric dust and water droplets and absorbed by carbon dioxide, ozone and water vapour molecules. Therefore, all radiation thermometers have to be carefully calibrated for each separate application.

Optical pyrometer

The optical pyrometer, illustrated in Figure 10.10, is designed to measure temperatures where the peak radiation emission is in the red part of the visible

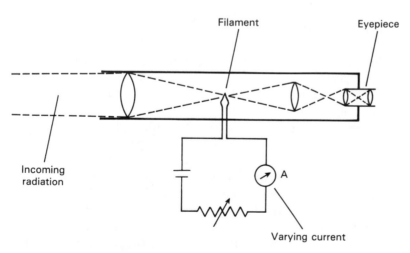

Figure 10.10 The optical pyrometer

spectrum, i.e. where the measured body glows a certain shade of red according to the temperature. This limits the instrument to measuring temperatures above 600 °C. The instrument contains a heated tungsten filament within its optical system. The current in the filament is increased until its colour is the same as the hot body; under these conditions the filament apparently disappears when viewed against the background of the hot body. Temperature measurement is therefore obtained in terms of the current flowing in the filament. As the brightness of different materials at any particular temperature varies according to the emissivity of the material, the calibration of the optical pyrometer must be adjusted according to the emissivity of the target. Manufacturers provide tables of standard material emissivities to assist with this.

The inherent measurement inaccuracy of an optical pyrometer is ±5 °C. However, in addition to this error, there can be a further operator-induced error of ±10 °C arising from the difficulty in judging the moment when the filament 'just' disappears. Measurement accuracy can be improved somewhat by employing an optical filter within the instrument which passes a narrow band of frequencies of wavelength around 0.65 μm corresponding to the red part of the visible spectrum.

The instrument cannot be used in automatic temperature control schemes because the eye of the human operator is an essential part of the measurement system. The reading is also affected by fumes in the sight path. Because of these difficulties and low accuracy, the optical pyrometer is rapidly being overtaken by the hand-held radiation pyrometer, although the optical pyrometer is still widely used in industry for measuring temperatures in furnaces and similar applications.

Radiation pyrometers

All the alternative forms of radiation pyrometer described below have an optical system which is similar to that in the optical pyrometer and focusses the energy emitted from the measured body. They differ, however, by omitting the filament and eyepiece and using instead an energy detector in the same focal plane as the eyepiece was, as shown in Figure 10.11. The radiation detector is either a thermal detector, which measures the temperature rise in a black body at the focal point of the optical system, or a photon detector. Thermal detectors respond equally to all wavelengths in the frequency spectrum, whereas photon detectors respond selectively to a particular band within the full spectrum.

Thermopiles, resistance thermometers and thermistors are all used as thermal detectors in different versions of these instruments. These typically have time constants of several milliseconds because of the time taken for the black body to heat up and the temperature measuring instrument to respond to the temperature change.

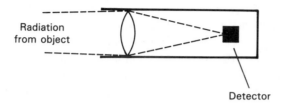

Radiation
from object

Detector

Figure 10.11 Structure of the radiation thermometer

Photon detectors are usually of the photoconductive or photovoltaic type, both of which respond very much faster to temperature changes than thermal detectors because they involve atomic processes, and typical measurement time constants are a few microseconds.

Various forms of electrical output are available from the radiation detector; these are functions of the incident energy on the detector and are therefore functions of the temperature of the measured body. While this makes such instruments useful in automatic control systems, their accuracy is often inferior to that of optical pyrometers. This reduced accuracy arises firstly because a radiation pyrometer is sensitive to a wider band of frequencies than the optical instrument, and the relationship between emitted energy and temperature is less well defined. Secondly, the magnitude of energy emission at low temperatures becomes very small, according to equation (10.5), increasing the difficulty of accurate measurement.

The various forms of radiation pyrometer differ mainly in the technique used to measure the emitted radiation. They also differ in the range of energy wavelengths which they are designed to measure and hence in the temperature range measured. One further difference is the material used to construct the energy-focussing lens. Outside the visible part of the spectrum, glass becomes almost opaque to infra-red wavelengths, and other lens materials, such as arsenic trisulphide, are used. Further information about these can be found elsewhere.[2]

Intelligent temperature measuring instruments

Intelligent temperature transmitters have recently been introduced into the catalogues of instrument manufacturers, and they bring about the usual benefits associated with intelligent instruments. Such transmitters are designed for use with transducers which have either a dc voltage output in the mV range or an output in the form of a resistance change. They are therefore suitable for use in conjunction with thermocouples, thermopiles, resistance thermometers, thermistors and broad-band radiation pyrometers. All the transmitters presently available have non-volatile memories where all constants used in correcting output values for modifying inputs, etc., are stored, thus enabling the instrument to survive power failures without losing such information.

The transmitters now available include adjustable damping, noise rejection, self-adjustment for zero and sensitivity drifts and expanded measurement range. These features allow an inaccuracy level of only ±0.05 per cent of full scale to be specified.

Intelligent instruments cost significantly more than their non-intelligent counterparts, and justification purely on the grounds of their superior accuracy is difficult. However, their expanded measurement range means immediate savings are made in terms of the reduction in the number of spare instruments needed to cover a number of measurement ranges. Their capability for self-diagnosis and self-adjustment means that they require attention much less frequently, giving additional savings in maintenance costs.

An intelligent temperature-measuring instrument should be properly regarded as a two-part measurement system consisting of a temperature transducer and a signal processing element. Although the signal processing element is largely self-calibrating, the appropriate calibration routines described earlier have to be applied to the transducer.

10.2 Introduction to calibration

The primary standard of temperature is fundamentally different in character from the primary standards of mass, length and time. If two bodies of lengths l_1 and l_2 are connected together end to end, the result is a body of length $l_1 + l_2$. A similar relationship exists between separate masses and separate times. However, if two bodies at the same temperature are connected together, the joined body has the same temperature as each of the original bodies.

This is a root cause of the fundamental difficulties which exist in establishing an absolute standard for temperature in the form of a relationship between it and other measurable quantities for which a primary standard unit exists.

One obvious way of defining temperatures in absolute units would seem to be to measure the heat flowing between two bodies at different temperatures. An attempt to do this was made by Kelvin in 1848. He proposed a thermodynamic temperature scale, based on the Carnot cycle, which defines the temperature ratio between two bodies A and B. If two heat reservoirs of infinite capacity, A and B, at temperatures T_A and T_B are put into contact, the heat flow from A, Q_A, and the heat flow to B, Q_B, are related by

$$\frac{T_A}{T_B} = \frac{Q_A}{Q_B}$$

Unfortunately, this is not realizable physically because it is based on an ideal Carnot cycle which cannot be achieved practically.

An identical temperature scale can be defined by a constant volume or constant pressure gas thermometer using an ideal gas, and we have for example,

$$\frac{T_1}{T_2} = \frac{P_1}{P_2} \text{ (from ideal gas law: } PV = RT)$$

Again, however, this is of little practical use since real gases only approach the behaviour of ideal gases as their pressure is reduced towards zero. The procedure which must be followed to make use of this relationship is to measure the pressure ratio at several low pressures and extrapolate the result to zero pressure. This would be a very time consuming procedure for normal calibration procedures and in any case the accuracy yielded by a gas thermometer is inadequate.

In the absence of a convenient relationship that relates the temperature of a body to another measurable quantity which can be expressed in primary standard units, it is necessary to establish fixed, reproducable reference points for temperature in the form of freezing and boiling points of substances where the transition between solid, liquid and gaseous states is sharply defined. The International Practical Temperature Scale (IPTS) uses this philosophy and defines six *primary fixed points* for reference temperatures in terms of

the triple point of equilibrium hydrogen	$-259.34\,^{\circ}\text{C}$
the boiling point of oxygen	$-182.962\,^{\circ}\text{C}$
the boiling point of water	$100.0\,^{\circ}\text{C}$
the freezing point of zinc	$419.58\,^{\circ}\text{C}$
the freezing point of silver	$961.93\,^{\circ}\text{C}$
the freezing point of gold	$1064.43\,^{\circ}\text{C}$

(all at standard atmospheric pressure). The thermodynamic temperatures of these states are established for calibration purposes by gas thermometers. For calibrating intermediate temperatures, interpolation between the fixed points is carried out by the following reference instruments:

(a) a platinum resistance thermometer in the temperature range
$-182.962\,^{\circ}\text{C}$ to $+630.74\,^{\circ}\text{C}$; and

(b) a platinum–platinum/10 per cent rhodium thermocouple in the temperature range $+630.74\,^{\circ}\text{C}$ to $+1064.43\,^{\circ}\text{C}$.

($630.74\,^{\circ}\text{C}$ is the melting temperature of the secondary fixed point metal antimony.)

Above the gold melting point, standard temperatures are defined by a narrow-band radiation thermometer using the formula

$$\frac{J_t}{J_G} = \frac{\exp(1.438/1336.15\lambda) - 1}{\exp(1.438/\lambda t) - 1}$$

where J_t is the spectral radiance of a black body at temperature t, J_G is the spectral radiance of the black body at the boiling point of gold, and λ is the wavelength of the pyrometer.

The freezing points of certain other metals are also used as *secondary fixed points* to provide additional reference points during calibration procedures. These are of particular use for calibrating instruments measuring high temperatures. Some examples are

freezing point of tin:	231.968 °C
freezing point of lead:	327.502 °C
freezing point of zinc:	419.58 °C
freezing point of antimony:	630.74 °C
freezing point of aluminium:	660.37 °C
freezing point of copper:	1084.5 °C
freezing point of nickel:	1455 °C
freezing point of palladium:	1554 °C
freezing point of platinum:	1772 °C
freezing point of rhodium:	1963 °C
freezing point of iridium:	2447 °C
freezing point of tungsten:	3387 °C

The accuracy of temperature calibration procedures is fundamentally dependent on how accurately points on the IPTS can be reproduced. The present limits are

1 K	0.3%	800 K	0.001%
10 K	0.1%	1500 K	0.02%
100 K	0.005%	4000 K	0.2%
273.15 K	0.0001%	10 000 K	6.7%

10.3 Calibration instruments and equipment

Almost any instrument from the list presented in Section 10.1 might be used for calibration duties in certain circumstances. Where involved in such duties, of course, the instrument used would be one of high accuracy which was reserved solely for calibration duties. The list of calibration instruments therefore includes mercury-in-glass thermometers, base-metal thermocouples (type K), noble-metal thermocouples (types B, R and S), platinum resistance thermometers and radiation pyrometers.

Although no special instruments have been developed for temperature calibration, certain purpose-designed equipment is required to provide the controlled temperatures necessary for the calibration process. Such equipment is commercially available from several sources.

For calibration of all temperature transducers other than radiation thermometers, a furnace consisting of an electrically heated ceramic tube is commonly used. The temperature of such a furnace can typically be controlled within limits of $\pm 2\,°C$ over the range from 20 °C to 1600 °C. Up to 630 °C, the platinum resistance thermometer is commonly used as a reference standard. Above that temperature, a type R (platinum/13 per cent rhodium–platinum) thermocouple is usually employed. Type K (chromel–alumel) thermocouples are also used as an alternative reference standard for temperature calibration up to 1000 °C.

Below 20 °C, a stirred water bath is used to provide a constant reference temperature, and the same equipment can be used for temperatures up to 100 °C. Similar stirred liquid baths containing oil or salts (potassium–sodium nitrate mixtures) can be used to provide reference temperatures up to 600 °C.

All the above equipment provides an environment in which calibration is performed by comparing one instrument with another. Absolute reference standards of temperature can only be obtained by the freezing point method in which an ingot of pure metal (better than 99.99 per cent pure), protected against oxidation inside a graphite crucible with a close fitting lid, is heated beyond its melting point and allowed to cool. If its temperature is monitored, an arrest period is observed in its cooling curve at the freezing point of the metal. Electric resistance furnaces are available to carry out this procedure. Up to 1100 °C, a measurement uncertainty of less than $\pm 0.5\,°C$ is achievable. It should be noted that this technique only provides standard reference temperatures at the fixed melting points of the metals used. However, this is still a very useful check on the accuracy of reference standard instruments, and, once their accuracy has been verified at such points within their measuring range, much greater confidence is created in their accuracy at intermediate temperatures.

For the calibration of radiation thermometers, a radiation source which approximates as closely as possible to the behaviour of a black body is required. The actual value of the emissivity of the source must be measured by a surface pyrometer. Some form of optical bench is also required so that instruments being calibrated can be held firmly and aligned accurately.

The simplest form of radiation source is a hot plate heated by an electrical element. The temperature of such devices can be controlled within limits of $\pm 1\,°C$ over the range from 0 °C to 650 °C and the typical emissivity of the plate surface is 0.85. Type R noble-metal thermocouples embedded in the plate are normally used as the reference instrument.

A black-body cavity provides a heat source with a much better emissivity. This can be constructed in various alternative forms according to the temperature range of the radiation thermometers to be calibrated, though a common feature is a blackened conical cavity with a cone angle of about 15°.

For calibrating low-temperature radiation pyrometers (measuring temperatures in the range 20 °C–200 °C), the black-body cavity is maintained at a constant temperature (±0.5 °C) by immersing it in a liquid bath. The typical emissivity of a cavity heated in this way is 0.995 . Water is suitable for the bath in the temperature range 20 °C–90 °C and a silicone fluid is suitable for the range 80 °C–200 °C. Within these temperature ranges, a mercury-in-glass thermometer is commonly used as the standard reference calibration instrument, although a platinum resistance thermometer is used when better accuracy is required.

Another form of black-body cavity is one lined with a refractory material and heated by an electrical element. This gives a typical emissivity of 0.998 and is used for calibrating radiation pyrometers at higher temperatures. Within the range 200 °C–1200 °C, temperatures can be controlled within limits of ±0.5 °C and a type R thermocouple is generally used as the reference instrument. At the higher range 600 °C–1600 °C, temperatures can be controlled within limits of ±1 °C and a type B thermocouple (30 per cent rhodium–platinum/6 per cent rhodium–platinum) is normally used as the reference instrument. As an alternative to thermocouples, radiation thermometers can also be used as a standard within ±0.5 °C over the temperature range 400 °C–1250 °C.

To provide reference temperatures above 1600 °C, a carbon cavity furnace is used. This consists of a graphite tube with a conical radiation cavity at its end. Temperatures up to 2600 °C can be maintained with an accuracy of ±5 °C. Narrow-band radiation thermometers are used as the reference standard instrument.

Again, the above equipment merely provides an environment in which radiation thermometers can be calibrated against some other reference standard instrument. To obtain an absolute reference standard of temperature, a fixed-point, black-body furnace is used. This has a radiation cavity consisting of a conical-ended cylinder which contains a crucible of 99.999 per cent pure metal. If the temperature of the metal is monitored as it is heated up at a constant rate, an arrest period is observed at the melting point of the metal where the temperature ceases to rise for a short time interval. Thus the melting point, and hence the temperature corresponding to the output reading of the monitoring instrument at that instant, are defined exactly. Measurement uncertainty is of the order of ±0.3 °C. The list of metals, and their melting points, used to provide such primary and secondary reference temperatures can be found in Section 10.2.

In the calibration of radiation thermometers, knowledge of the emissivity of the hot plate or black-body furnace used as the radiation source is essential. This is measured by special types of surface pyrometer. Such instruments contain a hemispherical, gold-plated surface which is supported on a telescopic arm that allows it to be put into contact with the hot surface. The radiation emitted from a small hole in the hemisphere is independent of the surface emissivity of the measured body and is equal to that which would be emitted by the body if its emissivity value were 100. This radiation is measured by a thermopile with its cold junction at a controlled temperature. A black hemisphere is also provided with the instrument which can be inserted to cover the gold surface. This allows the instrument to measure the normal radiation emission from the hot body and so allows the surface emissivity to be calculated by comparing the two radiation measurements.

Within this list of special equipment, mention must also be made of standard tungsten strip lamps which are used for providing constant known temperatures in the calibration of optical pyrometers. The various versions of these provide a range of standard temperatures between 800 °C and 2300 °C to an accuracy of ± 2 °C.

10.4 Calibration frequency

The manner in which the required frequency for calibration checks is determined for the various temperature-measuring instruments available was discussed in the instrument review presented in Section 10.1. The simplest instruments from a calibration point of view are liquid-in-glass thermometers. The only parameter able to change within these is the volume of the glass used in their construction. This only changes very slowly with time, and hence only infrequent (e.g. annual) calibration checks are required.

The required frequency for calibration of all other instruments is either (a) dependent upon the type of operating environment and the degree of exposure to it or (b) use-related. In some cases, both these factors are relevant.

Resistance thermometers and thermistors are examples of instruments where the drift in characteristics depends on the environment they are operated in and on the degree of protection they have from that environment. Devices such as gas thermometers and quartz thermometers suffer characteristics drift which is largely a function of how much they are used (or misused!), though in the case of quartz thermometers, any drift is likely to be small and only infrequent calibration checks will be required. Any instruments not mentioned so far suffer characteristics drift due to both environmental and use-related factors. The list of such instruments includes bimetallic thermometers, thermo-couples, thermopiles and radiation thermometers. In the case of thermocouples and thermopiles, it must be remembered that error in the required character-istics is possible even when the instruments are new, as discussed in Section 10.1, and therefore their calibration must be checked before use.

As the factors responsible for characteristics drift vary from application to application, the required frequency of calibration checks can only be determined experimentally. The procedure for this is to start by checking the calibration of instruments used in new applications at very short intervals of time, and then to lengthen the interval between calibration checks progressively until a significant deterioration in instrument characteristics is observed. The required calibration interval is then defined as that time interval which is predicted to elapse before the characteristics of the instrument have drifted to the limits which are allowable in that particular measurement application. A suggested mechanism for this procedure was described more fully in Section 9.4.

Working and reference standard instruments and ancillary equipment must also be calibrated periodically. An interval of two years is usually recommended between such calibration checks, although monthly checks are advised for the black-body-cavity furnaces used to provide standard reference temperatures in pyrometer calibration. Standard resistance thermometers and thermocouples may also need more frequent calibration checks if the conditions (especially of temperature) and frequency of use demand them.

10.5 Calibration procedures

The standard way of calibrating temperature transducers is to place them into a temperature controlled environment together with a standard instrument,[3,4] or to use a radiant heat source of controlled temperature with high emissivity in the case of radiation thermometers. In either case, the controlled temperature must be measured by a standard instrument whose calibration is traceable to reference standards. This is a suitable method for most instruments in the calibration chain but is not necessarily appropriate or even possible for the process instruments at the lower end of the chain.

In the case of many process instruments, their location and mode of fixing makes it difficult or sometimes impossible to remove them to a laboratory for calibration checks to be carried out. In this event, it is standard practice to cali-brate them in their normal operational position, using a reference instrument which is able to withstand whatever hostile environment may be present. If this practice is followed, it is imperative that the working standard instrument is checked regularly to ensure that it has not been contaminated.

Such *in-situ* calibration may also be required where process instruments have characteristics which are sensitive to the environment in which they work, so that they are calibrated under their usual operating conditions and are therefore accurate in normal use. However, the preferred way of dealing with this situation is to calibrate them in a laboratory with ambient conditions (of pressure, humidity, etc.) set up to mirror those of the normal operating environment. This alternative avoids having to subject reference calibration instruments to harsh chemical environments which are commonly associated with manufacturing processes.

For instruments at the lower end of the calibration chain, i.e. those measuring process variables, it is common practice to calibrate them against an instrument which is of the same type but of higher accuracy and reserved only for calibration duties. If a large number of different types of instrument have to be calibrated, however, this practice leads to the need to keep a large number of different calibration instruments. To avoid this, various reference instruments are available which can be used to calibrate all process instruments within a given temperature-measuring range. Examples are the liquid-in-glass thermometer (0 °C–200 °C), platinum resistance thermometer (− 200 °C–630 °C) and type B thermocouple (600 °C–1600 °C). The optical pyrometer is also often used as a reference instrument at this level for the calibration of other types of radiation thermometers.

For calibrating instruments further up the calibration chain, particular care is needed with regard to both the instruments used and the conditions they are used under. It is difficult and expensive to meet these conditions and hence this function is sub-contracted by most companies to specialist laboratories. The reference instruments used are the platinium resistance thermometer in the temperature range − 200 °C to + 630 °C, the platinum–platinum/10 per cent rhodium (type S) thermocouple in the temperature range + 630 °C to + 1064 °C and a narrow band radiation thermometer at higher temperatures. An exception is optical pyrometers, which are calibrated by sighting them on the filament of a calibrated tungsten strip lamp in which the current is accurately measured, providing calibration at temperatures up to 2500 °C. A particular note of caution must be made where platinum–rhodium thermocouples are used as a standard. These are very prone to contamination and, if they need to be handled at all, this should be done with very clean hands.

Before ending this chapter, it is appropriate to mention one or two further points which concern the calibration of thermocouples and radiation thermometers.

As mentioned earlier, apart from calibration checks in the normal way, the calibration of even new *thermocouples* must be checked before use. The procedure for this is to immerse both junctions of the thermocouple in an ice-bath and measure its output with a high-accuracy digital voltmeter (± 5 μV). Any output greater than 5 μV would indicate a fault in the thermocouple material and/or its construction. When carrying out any calibration operations on thermocouples, it is important to exclude any source of electrical or magnetic fields, as these will induce erroneous voltages in the sensor.

Special comments are also relevant regarding the calibration of a *radiation thermometer*. As well as the normal accuracy checks, its long-term stability must also be verified by testing its output over a period which is one hour longer than the manufacturer's specified 'warm-up' time. This shows up any components within the instrument which are suffering from temperature-induced characteristics drift. It is also necessary to calibrate radiation thermometers according to the emittance characteristic of the body whose temperature is being measured and according to the level of energy losses in the radiation path between the body and measuring instrument. Such emissivity calibration must be carried out for every separate application that the instrument is used for, using a surface pyrometer.

Optical pyrometers are calibrated by sighting them on the filament of a tungsten strip lamp in which the current is accurately measured. This method of calibration can be used at temperatures up to 2500 °C. Alternatively, they can be calibrated against a standard radiation pyrometer.

Pressure calibration

11.1 Review of pressure-measuring instruments

Pressure-measuring instruments normally measure gauge pressure rather than absolute pressure because the measurement of absolute pressure is very difficult. The exception to this is one form of U-tube manometer which does measure absolute pressure, although with limited accuracy.

The instruments available for measuring pressures can be divided into four categories according to the range of pressure which they are designed to measure. These four pressure ranges are as follows:

(a) high pressures greater than 7000 bar (6910 atmospheres);
(b) mid-range pressures in the span between 1.013 bar and 7000 bar (1–6910 atmospheres);
(c) low pressures between 0.001 mbar and 1.013 bar (1 atmosphere);
(d) very low pressures less than 0.001 mbar.

Apart from certain types of diaphragm-type instrument which use either a piezo-electric crystal or a capacitive transducer to measure the pressure-induced displacement, all the instruments discussed in this chapter are restricted to measuring only static pressures. Instruments and techniques for measuring dynamic pressures have not been included because the measurement of such is a very specialized area which is not of general interest.

Measurement of high pressures above 7000 bar is normally carried out electrically by monitoring the change of resistance of wires of special materials. Materials having resistance–pressure characteristics which are suitably linear and sensitive include gold–chrome alloys and manganin.

Mid-range pressures are the ones measured most commonly in industrial applications. Suitable instruments for this range are the U-tube manometer, the well-type manometer, the inclined manometer, the diaphragm, Bourdon tubes, bellows-type instruments and resonant wire devices.

Low pressures are measurable by special versions of the common mid-range pressure-measuring instruments. Suitable instruments include special manometers, Bourdon tubes, bellows-type instruments and diaphragms.

For the very low-pressure range, special instruments have been developed. These include the thermocouple gauge, Pirani gauge, thermistor gauge and ionization gauge.

The mode of operation and characteristics of these various pressure-measuring devices are considered below. More detailed coverage is given elsewhere.[1]

U-tube manometers

All U-tube manometers consist of a glass vessel shaped into a letter U and filled with a liquid. Two alternative forms of the instrument exist.

In one form, shown in Figure 11.1, one end of the U-tube is sealed and evacuated. This is the form mentioned earlier which measures absolute pressure. The absolute pressure is measured in terms of the difference between the mercury levels in the two halves of the tube. Thus

$$p_1 \text{(absolute pressure)} = h\rho$$

Quite apart from the difficulty in judging exactly where the meniscus levels in the mercury are, such an instrument cannot give a perfect measurement because of the impossibility of achieving a total vacuum at the sealed end of the tube. Although it is possible by modern techniques to design an instrument which does give a reasonably accurate measurement of absolute pressure, the problem is usually avoided in practice by measuring gauge pressure instead of absolute pressure.

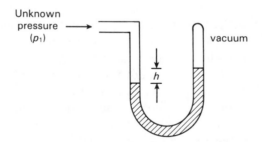

Figure 11.1 Sealed U-tube measuring absolute pressure

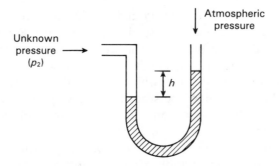

Figure 11.2 U-tube measuring gauge pressure

The alternative form of U-tube manometer is the one shown in Figure 11.2, where both ends of the tube are open. If an unknown pressure is applied to one end and the other end is left open to the atmosphere, gauge pressure is measured. The unknown gauge pressure of the fluid (p_1) is related to the difference (h) in the levels of fluid in the two halves of the tube by the expression

$$p_2(\text{gauge pressure}) = h\rho$$

where ρ is the specific gravity of the fluid.

The form of U-tube manometer with two open ends can also be used to measure differential pressure, as shown in Figure 11.3. The unknown pressures (p_3 and p_4) are applied to the ends of the tube, and the differential pressure ($p_3 - p_4$) is then given by

$$(p_3 - p_4) = h\rho$$

The type of liquid used in the instrument depends on the pressure and characteristics of the fluid being measured. Water is a convenient and certainly

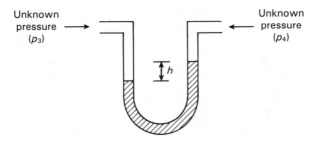

Figure 11.3 U-tube measuring differential pressure

cheap choice, but it evaporates easily and is difficult to see. It is nevertheless used extensively, with the major obstacles to its use being overcome by using coloured water and regularly topping up the tube to counteract evaporation.

Situations where water is definitely not used as a U-tube manometer fluid include the measurement of fluids which react with or dissolve in water, and also where higher-magnitude pressure measurements are required. In such circumstances, liquids such as aniline, carbon tetrachloride, bromoform, mercury or transformer oil are used.

U-tube manometers are normally used to measure mid-range fluid pressures up to 2 bar in magnitude. However, special versions exist which are designed to measure low pressures down to 0.1 mbar.

As far as calibration requirements are concerned, there are very few reasons for the measurement characteristics of U-tube manometers to drift. In the longer term, however, small errors can be introduced through volumetric changes in the glass. Annual calibration checks are thus advisable.

Well-type manometer (cistern manometer)

The well-type manometer, shown in Figure 11.4, is effectively a U-tube manometer in which one-half of the tube is made very large and takes the form of a well. The change in the level of the well as the measured pressure varies is negligible. Therefore, the liquid level in only one tube has to be measured, which makes the instrument much easier to use than the U-tube manometer. Gauge pressure is given by

$$p_1 = h\rho$$

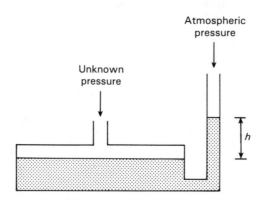

Figure 11.4 Well-type manometer

It might appear that the instrument would give a better measurement accuracy than the U-tube manometer because the need to subtract two liquid-level measurements in order to arrive at the pressure value is avoided. However, this benefit is swamped by errors which arise due to the typical cross-sectional area variations in the glass used to make the tube. Such variations do not affect the accuracy of the U-tube manometer. Annual calibration checks are advised as for the U-tube manometer.

Inclined manometer (draft gauge)

The inclined manometer, shown in Figure 11.5, is a variation on the well-type manometer in which one leg of the tube is inclined to increase measurement sensitivity. Similar comments to those above apply to accuracy and calibration requirements.

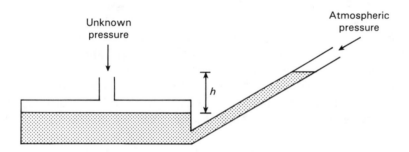

Figure 11.5 Inclined manometer

Diaphragm

The diaphragm is one of the three common types of elastic-element pressure transducer, and is shown schematically in Figure 11.6. Applied pressure causes displacement of the diaphragm and this movement is measured by a displacement transducer. Both gauge pressure and differential pressure can be measured by different versions of diaphragm-type instruments. In the case of differential pressure, the two pressures are applied to either side of the diaphragm and the displacement of the diaphragm corresponds to the pressure difference. The typical magnitude of displacement in either version is 0.1 mm, which is well suited to a strain-gauge type of measuring transducer.

Four strain gauges are normally used in a bridge configuration, in which an excitation voltage is applied across two opposite points of the bridge. The

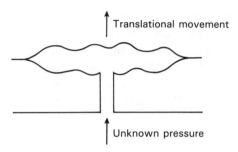

Figure 11.6 Diaphragm

output voltage measured across the other two points of the bridge is then a function of the resistance change due to the strain in the diaphragm. This arrangement automatically provides compensation for environmental temperature changes. Older pressure transducers of this type used metallic strain gauges bonded to a diaphragm typically made of stainless steel. Apart from manufacturing difficulties arising from the problem of bonding the gauges, metallic strain gauges have a low gauge factor. This means that the strain-gauge bridge only has a low output which has to be amplified by an expensive dc amplifier. The development of semi-conductor (piezo-resistive) strain gauges provided a solution to the low-output problem, as they have gauge factors up to one hundred times greater than metallic gauges. However the difficulty of bonding gauges to the diaphragm remained and a new problem emerged regarding the highly non-linear characteristic of the strain–output relationship.

The problem of strain-gauge bonding was solved with the emergence of monolithic piezo-resistive pressure transducers, and these are now the most commonly used type of diaphragm pressure transducer. The monolithic cell consists of a diaphragm made of a silicon sheet into which resistors are diffused during the manufacturing process. Besides avoiding the difficulty with bonding, such monolithic silicon measuring cells have the advantage of being very cheap to manufacture in large quantities and their sensitivity is three times better than that of unbonded strain gauges. Another advantage is that they can be made extremely small, down to 0.75 mm in diameter, and in this form find application in specialist areas such as on the tip of medical catheters. Although the inconvenience of a non-linear characteristic remains, this is normally overcome by processing the output signal with an active linearization circuit or incorporating the cell into a microprocessor-based intelligent measuring transducer. The latter usually provides analogue-to-digital conversion and interrupt facilities within a single chip and gives a digital output which is readily integrated into computer control schemes. Such instruments can also offer automatic temperature compensation, built-in diagnostics, and simple calibration procedures. These features allow measurement inaccuracies as low as ±0.1 per cent of full-scale reading.

As an alternative to strain-gauge-type displacement measurement, capacitive or inductive transducers are sometimes used. As a further option, developments in optical fibres have been exploited in the Fotonic sensor (Figure 11.7), which is a diaphragm-type device in which the displacement is measured by optoelectronic means. In the device, light travels from a light source down an optical fibre, is reflected back from the diaphragm and travels back along a second fibre to a photodetector. There is a characteristic relationship between the light reflected and the distance from the fibre ends to the diaphragm, thus making the amount of reflected light dependent upon the measured pressure.

Normally, diaphragm-type instruments are used to measure mid-range gauge and differential pressures in the range up to 10 bar. However, special versions are available which can measure low pressures down to 0.001 mbar.

While the discussion so far has been about measuring static pressures, diaphragm-type instruments can also be used to measure *dynamic pressures*. The principal displacement-measuring device used in such circumstances is the piezo-electric crystal. However, the strain gauge in its various forms can be used as an alternative, and there are particular advantages in its use in respect of its greater measurement sensitivity and also its ability to measure static pressures. A further alternative for measuring dynamic pressures is the diaphragm-based capacitive displacement transducer.

As diaphragms have mechanical components, their measurement characteristics can be affected by their operating environment and also by mishandling. As the magnitude of this effect is use-related, the necessary interval between re-calibration must be determined by practical experimentation.

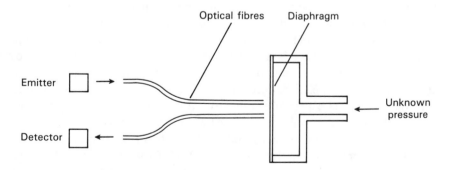

Figure 11.7 Fotonic sensor

Bellows

The bellows, illustrated in Figure 11.8, is another elastic-element-type device. It operates on a very similar principle to the diaphragm, although its use is much less common. The sensitivity of a bellows is greater than that of a diaphragm, which is its principal attribute. Pressure changes within the bellows produce translational motion of the end of the bellows which can be measured by capacitive, inductive (LVDT) or potentiometric transducers according to the range of movement produced. A typical measuring range for a bellows-type instrument is 0–1 bar (gauge pressure). However, special versions are available which are designed to measure low pressures down to 0.1 mbar.

As already stated, bellows are very similar to diaphragms in their operational characteristics and the required calibration frequency must be determined in the same manner.

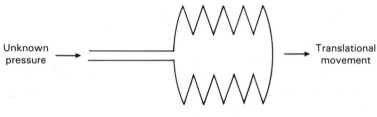

Figure 11.8 Bellows

Bourdon tube

The Bourdon tube is the third type of elastic-element pressure transducer and is a very common industrial measuring instrument which is used for measuring the pressure of both gaseous and liquid fluids. It consists of a specially shaped piece of oval-section flexible tube which is fixed at one end and free to move at the other. When pressure is applied at the fixed end of the tube, the oval cross-section becomes more circular. As the cross-section of the tube tends towards a circular shape, a deflection of the closed, free end of the tube is caused. When used as a pressure indicator, this motion is translated into the movement of a pointer against a scale. Alternatively, if an electrical output is required, the displacement is measured by some form of displacement transducer, which is commonly a potentiometer or LVDT (linear variable differential transformer), or less often a capacitive sensor. In yet another version, the displacement is measured optically.

The three common shapes of Bourdon tube are shown in Figure 11.9(a),

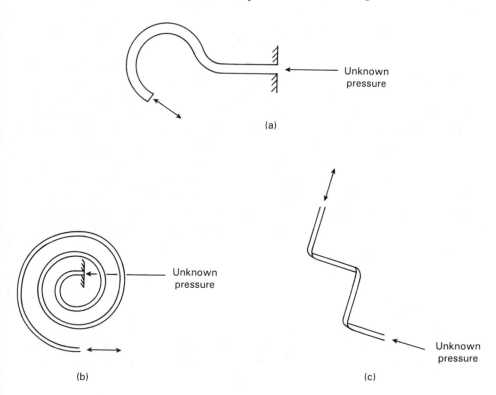

(a)

(b)

(c)

Figure 11.9 Bourdon tubes: (a) C-type; (b) spiral type; (c) helical type

(b) and (c). The maximum possible deflection of the free end of the tube is proportional to the angle subtended by the arc through which the tube is bent. For a C-type tube, the maximum value for this arc is somewhat less than 360°. Where greater measurement sensitivity and resolution are required, spiral and helical tubes are used in which the possible magnitude of the arc subtended is limited only by a practical limit on how many turns it is convenient to have in the helix or spiral. However, this increased measurement performance is only gained at the expense of a substantial increase in manufacturing difficulty and cost compared with C-type tubes, and is also associated with a large decrease in the maximum pressure which can be measured.

C-type tubes are available for measuring pressures up to 6000 bar. A typical C-type tube of 25 mm radius has a maximum displacement travel of 4 mm, giving a moderate level of measurement resolution. Measurement inaccuracy is typically quoted as ±1 per cent of full-scale deflection. Similar accuracy is available from helical and spiral types, but while the measurement resolution is higher, the maximum pressure measurable is only 700 bar. Special versions of the Bourdon tube are also available for measuring low pressures in the range down to 10 mbar.

Because both the Bourdon tube itself and the secondary transducer used to measure the deflection of its end are mechanical in nature, the amount of the change in characteristics with time is dependent upon the operational conditions prevailing. The re-calibration frequency necessary must therefore be determined experimentally. Mention should also be made of the practical difficulties of calibrating Bourdon tubes which are to be used for measuring liquid pressures, as discussed more fully later in Section 11.5. In consequence of this, the accuracy limits quoted earlier are only guaranteed when gaseous pressures are being measured. It is possible to achieve these same levels of accuracy when measuring liquid pressures, but only if the gauge can be totally filled with liquid during both calibration and measurement, a condition which is very difficult to fulfil practically.

Resonant wire devices

The resonant wire device is a relatively new instrument which has emanated from recent advances in the electronics field. A typical device is shown schematically in Figure 11.10. Wire is stretched across a chamber containing fluid

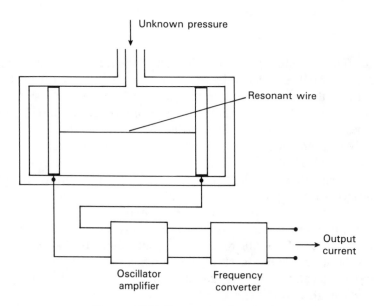

Figure 11.10 Resonant wire device

at unknown pressure subjected to a magnetic field. The wire resonates at its natural frequency according to its tension, which varies with pressure. Thus pressure is calculated by measuring the frequency of vibration of the wire using electronics integrated into the cell. Such devices are highly accurate, with an inaccuracy of ±0.2 per cent of full-scale reading being typical, and they are particularly insensitive to ambient condition changes.

Because of their newness, little practical experience has been gained in the long-term operation of these devices. Comment about the required re-calibration frequency must therefore be a little guarded. However, there is little within the instrument which seems likely to suffer characteristics-drift and therefore the initial recommendation would be to carry out calibration checks annually.

Thermocouple gauge

The thermocouple gauge is one of the group of gauges working on the thermal conductivity principle. The Pirani and thermistor gauges also belong to this group. At low pressure, the kinematic theory of gases predicts a linear relationship between pressure and thermal conductivity. Thus measurement of thermal conductivity gives an indication of pressure.

Figure 11.11 shows a sketch of a thermocouple gauge. Operation of the gauge depends on the thermal conduction of heat between a thin, hot metal strip in the centre and the cold outer surface of a glass tube (which is normally

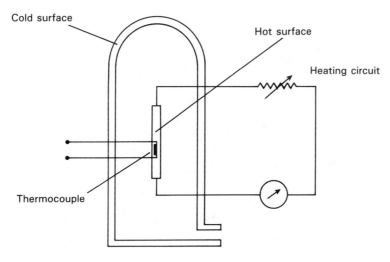

Figure 11.11 Thermocouple gauge

at room temerature). The metal strip is heated by passing a current through it and its temperature is measured by a thermocouple. The temperature measured depends on the thermal conductivity of the gas in the tube and hence on its pressure. A source of error in this instrument is the fact that heat is also transferred by radiation as well as conduction. This is of a constant magnitude, independent of pressure, and so could be measured and corrected for. However, it is usually more convenient to design for low radiation loss by choosing a heated element with low emissivity. Thermocouple gauges are typically used to measure pressures in the range 10^{-4} mbar up to 1 mbar.

The characteristics of the gauge vary with the nature of the gas whose pressure is being measured. Therefore, it must be calibrated separately for every application.

Pirani gauge

A typical form of Pirani gauge is shown in Figure 11.12. This is similar to a thermocouple gauge but has a heated element which consists of four coiled tungsten wires connected in parallel. Two identical tubes are normally used, connected in a bridge circuit as shown in Figure 11.13, with one containing the gas at unknown pressure and the other evacuated to a very low pressure. Current is passed through the tungsten element, which attains a certain temperature according to the thermal conductivity of the gas. The resistance of the

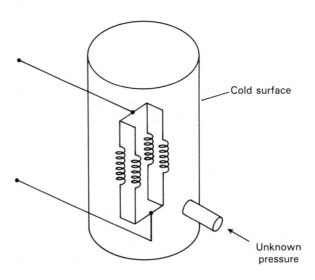

Figure 11.12 Pirani gauge

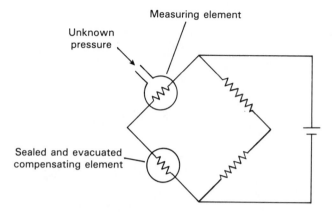

Figure 11.13 Wheatstone bridge circuit to measure output of Pirani gauge

element changes with temperature and causes an imbalance of the measurement bridge. Thus the Pirani gauge avoids the use of a thermocouple to measure temperature (as in the thermocouple gauge) by effectively using a resistance thermometer as the heated element. Such gauges cover the pressure range 10^{-5} mbar to 1 mbar.

As for the thermocouple gauge, the Pirani gauge must be calibrated separately for each different application.

Thermistor gauge

Thermistor gauges operate on identical principles to the Pirani gauge but use semiconductor materials for the heated elements instead of metals. The normal pressure range covered is 10^{-4} mbar to 1 mbar. Again, the gauge must be calibrated separately for each different application.

Ionization gauge

The ionization gauge is a special type of instrument used for measuring very low pressures in the range 10^{-13} to 10^{-3} bar. Gas of unknown pressure is introduced into a glass vessel containing free electrons discharged from a heated filament, as shown in Figure 11.14. Gas pressure is determined by measuring the current flowing between an anode–cathode system within the vessel. This current is proportional to the number of ions per unit volume, which in turn is proportional to the gas pressure. Measurement uncertainty varies between ±1 per cent at

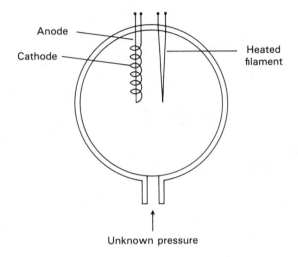

Figure 11.14 Ionization gauge

the top end of the measurement range to ±4 per cent at the lower end. No characteristics drift is anticipated in the instrument but calibration checks should nevertheless be carried out annually.

High-pressure measuring devices

The normal instrument used for measuring high pressures consists of a coil of manganin wire enclosed in a sealed, kerosene-filled, flexible bellows, as shown in Figure 11.15. The unknown pressure is applied to one end of the bellows

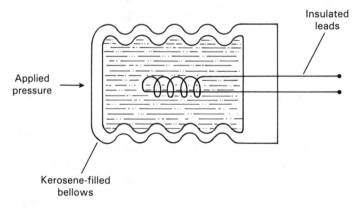

Figure 11.15 High-pressure measurement – wire coil in bellows

which transmits the pressure to the coil. The resistance is linearly proportional to pressure and therefore the magnitude of the applied pressure can be determined by measuring the coil resistance. Typical measurement inaccuracy is ±0.5 per cent.

Wire made from a gold–chrome alloy is sometimes used in place of manganin wire. This has a lower temperature coefficient but also a lower pressure sensitivity.

Intelligent pressure transducers

Adding microprocessor power to pressure transducers brings about substantial improvements in their characteristics. Measurement sensitivity improvement,

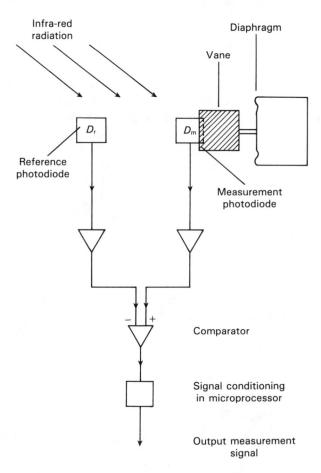

Figure 11.16 One type of intelligent pressure-measuring instrument

extended measurement range, compensation for hysteresis and other non-linearities, and correction for ambient temperature and pressure changes are just some of the facilities offered by intelligent pressure transducers. Inaccuracies of only ±0.1 per cent can be achieved with piezo-resistive-bridge silicon devices for instance.

Some recent microprocessor-based pressure transducers make use of novel techniques of displacement measurement. For example, both diaphragm and helical Bourdon tube devices are now available which use an optical method of displacement measurement of the form shown in Figure 11.16. In this, the motion is transmitted to a vane which progressively shades one of two monolithic photodiodes that are exposed to infra-red radiation. The second photodiode acts as a reference, enabling the microprocessor to compute a ratio signal which is linearized and is available as either an analogue or digital measurement of pressure. The typical measurement inaccuracy is ±0.1 per cent.

11.2 Introduction to calibration

Pressure is a quantity which is derived from the fundamental quantities of force and area, and is usually measured in terms of the force acting on a known area. In the mid-range of pressures from 0.1 mbar to 20 bar, U-tube manometers, dead-weight gauges and barometers are used to calibrate pressure-measuring instruments. The recently developed vibrating cylinder gauge also provides a very accurate reference standard over part of this range. Above 20 bar, a gold–chrome alloy resistance instrument is normally used. For low pressures in the range of 10^{-1} to 10^{-3} mbar, both the McLeod gauge and various forms of micromanometer are used as a pressure-measuring standard. At even lower pressures, below 10^{-3} mbar, a pressure-dividing technique is used to establish calibration. This involves setting up a series of orifices of accurately known pressure-ratio and measuring the upstream pressure with a McLeod gauge or micromanometer.

The limits of accuracy with which pressure can be measured by presently known techniques are as follows:

10^{-7} mbar	±4%
10^{-5} mbar	±2%
10^{-3} mbar	±1%
10^{-1} mbar	±0.1%
1 bar	±0.001%
10^4 bar	±0.1%

11.3 Pressure calibration instruments

Dead-weight gauge (pressure balance)

The dead-weight gauge or pressure balance, as shown in Figure 11.17, is a null-reading type of measuring instrument in which weights are added to the piston platform until the piston is adjacent to a fixed reference mark, at which time the downward force of the weights on top of the piston is balanced by the pressure exerted by the fluid beneath the piston. The fluid pressure is therefore calculated in terms of the weight added to the platform and the known area of the piston. The device is tedious to use but finds wide application as a reference instrument against which other pressure-measuring instruments are calibrated in the mid-range of pressures.

Special precautions are necessary in the manufacture and use of dead-weight gauges. Friction between the piston and cylinder must be reduced to a very low level, otherwise a significant measurement error would result. Friction reduction is accomplished by designing for a small clearance gap between the piston and cylinder, by machining the cylinder to a slightly greater diameter than the piston. The piston and cylinder are also designed so that they can be turned relative to one another which reduces friction still further. Unfortunately, as a result of the small gap between the piston and cylinder, there is a finite flow of fluid past the seals. This produces a viscous shear force which partly balances the dead-weight on the platform. A theoretical formula exists for calculating the magnitude of this shear force, suggesting that exact correction can be made for it. In practice, however, the piston deforms under pressure and alters the piston–cylinder gap, and so the shear force calculation and correction can only be approximate.

In spite of these difficulties, the instrument gives a typical measurement inaccuracy of only ±0.01 per cent. It is normally used for calibrating pressures

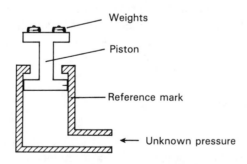

Figure 11.17 Dead-weight gauge

in the range of 20 mbar up to 20 bar. However, special versions can measure pressures down to 0.1 mbar or up to 7000 bar.

U-tube manometer

Along with its use for normal process measurements, the U-tube manometer is also used as a reference instrument for calibrating instruments measuring mid-range pressures. Although it is a deflection- rather than null-type instrument, it manages to achieve similar degrees of measurement accuracy to the dead-weight gauge, because of the error sources noted in the latter. The major source of error in U-tube manometers arises out of the difficulty in estimating the meniscus level of the liquid column accurately. There is also a tendency for the liquid level to creep up the tube by capillary action which creates an additional source of error.

U-tubes for measuring high pressures become unwieldy because of the long lengths of liquid-column and tube required. Consequently, U-tube manometers are normally used only for calibrating pressures at the lower end of the mid-pressure range.

Barometers

The most commonly used device type of barometer for calibration duties is the Fortin barometer. This is a highly accurate instrument which provides measurement inaccuracy levels of between ±0.03 per cent of full-scale reading and ±0.001 per cent of full-scale reading depending on the measurement range. To achieve such levels of accuracy, the instrument has to be used under very carefully controlled conditions of lighting, temperature and vertical alignment. It must also be manufactured to exacting standards and is therefore very expensive to buy. Corrections have to be made to the output reading according to the ambient temperature, local value of gravity and atmospheric pressure. Because of its expense and the difficulties in using it, the barometer is not normally used for calibration other than as a primary reference standard at the top of the calibration chain.

Vibrating cylinder gauge

The vibrating cylinder gauge, shown in Figure 11.18, acts as a reference standard instrument for calibrating pressure measurements up to 3.5 bar. It consists of a cylinder in which vibrations at the resonant frequency are excited by a

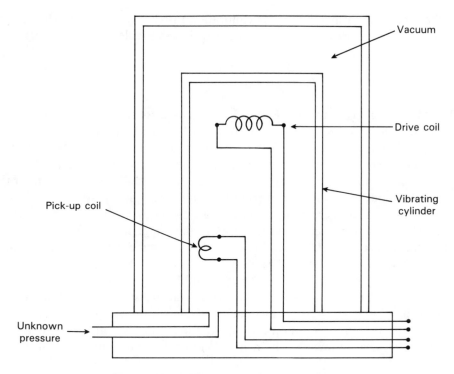

Figure 11.18 Vibrating cylinder gauge

current-carrying coil. The pressure-dependent oscillation frequency is monitored by a pick-up coil, and this frequency measurement is converted to a voltage signal by a microprocessor and signal conditioning circuitry contained within the package. By evacuating the space on the outer side of the cylinder, the instrument is able to measure the absolute pressure of the fluid inside the cylinder. Measurement errors are less than 0.005 per cent over the absolute pressure range up to 3.5 bar.

Gold–chrome alloy resistance instruments

For measuring pressures above 7000 bar, an instrument based on measuring the resistance change of a metal coil as the pressure varies is used, and the same type of instrument is also used for calibration purposes. As stated in Section 11.1, such instruments can either use manganin or gold–chrome alloys for the coil. Gold–chrome has a significantly lower temperature coefficient (i.e. its pressure–resistance characteristic is less affected by temperature changes), and is therefore the normal choice for calibration instruments in spite of its higher cost. An inaccuracy of only ± 0.1 per cent is achievable in such devices.

McLeod gauge

The McLeod gauge, shown in Figure 11.19, is used for the calibration of instruments designed to measure pressures below 0.1 mbar. Its principle of operation is to compress low-resistance fluid to a higher pressure which can be measured by manometer techniques. In essence, the gauge can be visualized as a U-tube manometer containing mercury, sealed at one end, and where the bottom of the U can be blocked at will. To operate the gauge, the piston is first withdrawn, causing the level of mercury in the lower part of the gauge to fall below the level of the junction J between the two tubes in the gauge marked Y and Z. Fluid at unknown pressure P_u is then introduced via the tube marked Z, from where it also flows into the tube marked Y, of cross-sectional area A. Next, the piston is pushed in, moving the mercury level up to block the junction J. At the stage where J is just blocked, the fluid in tube Y is at pressure P_u and contained in a known volume V_u. Further movement of the piston compresses the fluid in tube Y and this process continues until the mercury level in tube Z reaches a zero mark. Measurement of the height (h) above the mercury column in tube

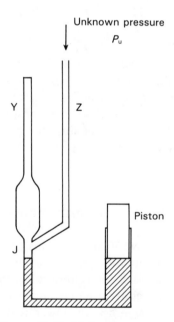

Figure 11.19 McLeod gauge

Y then allows calculation of the unknown pressure according to

$$P_u = \frac{A \cdot h^2 \cdot \rho \cdot g}{V_u - A \cdot h} \tag{11.1}$$

where ρ is the mass density of mercury. The compressed volume V_c is often very much smaller than the original volume, in which case equation (11.1) approximates to

$$P_u = \frac{A \cdot h^2 \cdot \rho \cdot g}{V_u} \text{ (for } A \cdot h \ll V_u) \tag{11.2}$$

Although the lowest measurement uncertainty achievable with McLeod gauges is ±1 per cent, this is still better than that which other gauges can achieve in the low-pressure range.

The McLeod gauge can measure low pressures directly down to 10^{-4} mbar. To measure lower pressures, it is necessary to use pressure-dividing techniques as described in Section 11.2.

Ionization gauge

The ionization gauge can calibrate instruments measuring low pressures across their whole range, and is therefore preferable to the McLeod gauge which can only calibrate instruments at specific points within their measuring range. The ionization gauge has a straight-line relationship between output reading and pressure. Unfortunately, its inherent accuracy is inadequate and it is only usable as a reference instrument if specific points on its output characteristic are calibrated against a McLeod gauge.

Micromanometers

Micromanometers are instruments which work on the manometer principle but are specially designed to minimize capillary effects and meniscus reading errors. The type of micromanometer which is most accurate as a calibration standard down to pressures of 10^{-3} mbar is the centrifugal micromanometer shown schematically in Figure 11.20. In this, a rotating disc serves to amplify a reference pressure, with the speed of rotation being adjusted until the amplified pressure just balances the unknown pressure. This null position is detected by observing when oil droplets sprayed into a glass chamber cease to move. Measurement accuracy is ±1 per cent.

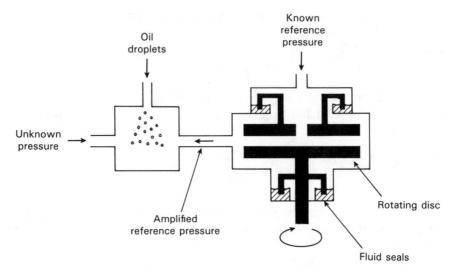

Figure 11.20 Centrifugal manometer

Other types of micromanometer also exist,[1] which give similar levels of accuracy, but only at somewhat higher pressure levels. These can be used as calibration standards at pressures up to 50 mbar.

11.4 Calibration frequency

Some pressure-measuring instruments are very stable and unlikely to suffer drift in characteristics with time. Devices in this class include resonant wire devices, ionization gauges and high-pressure instruments (working on the principle of resistance change with pressure). All forms of manometer are similarly stable, although in the longer term, small errors can develop in these through volumetric changes in the glass. Therefore, for all these instruments, only annual calibration checks are recommended, unless of course something happens to the instrument which puts its calibration into question.

However, most instruments used consist of an elastic element and a displacement transducer which measures its movement. Both of these component parts are mechanical in nature. Devices of this type include diaphragms, bellows and Bourdon tubes. Such instruments can suffer changes in characteristics for a number of reasons, for instance the characteristics of the operating environment and the degree to which the instrument is exposed to it. Another factor is the amount of mishandling it receives. These parameters are entirely dependent upon the particular application the instrument is used in, and the frequency with which it is used and exposed to the operating environment. A suitable calibration frequency can therefore only be determined on an experimental basis.

A third class of instrument from the calibration-requirements viewpoint is the range of devices that work on the thermal conductivity principle. This range includes the thermocouple gauge, Pirani gauge and thermistor gauge. Such instruments have characteristics which vary with the nature of the gas being measured and must therefore be calibrated each time they are used.

11.5 Calibration procedures

Pressure calibration requires the output reading of the instrument being calibrated to be compared with the output reading of a reference standard instrument when the same pressure is applied to both. This necessitates designing a suitable leakproof seal to connect the pressure-measuring chambers of the two instruments.

The calibration of pressure transducers used for process measurements often has to be carried out *in-situ* in order to avoid serious production delays. Such devices are often remote from the nearest calibration laboratory and to transport them there for calibration would take an unacceptably long time. Because of this, portable reference instruments have been developed for calibration at this level in the calibration chain. These use a standard air supply connected to an accurate pressure regulator to provide a range of reference pressures. An accuracy of ± 0.025 per cent is obtained when calibrating mid-range pressures in this manner. Calibration at higher levels in the calibration chain must of course be carried out in a proper calibration laboratory maintained in the correct manner. Irrespective of where calibration is carried out, however, several special precautions are necessary when using the various instruments described in Section 11.3.

U-tube manometers must have their vertical alignment carefully set up before use. Particular care must also be taken to ensure that there are no temperature gradients across the two halves of the tube. Such temperature differences would cause local variations in the specific weight of the manometer fluid, resulting in measurement errors. Correction must also be made for the local value of g (acceleration due to gravity). These comments apply similarly to the use of other types of manometer and micromanometer.

The existence of one potentially major source of error in *Bourdon tube* pressure measurement has not been widely documented and few manufacturers of Bourdon tubes make any attempt to so warn users of their products. The problem is concerned with the relationship between the fluid being measured and the fluid used for calibration. The pointer of the Bourdon tube is normally set at zero during manufacture, using air as the calibration medium. If, however, a different fluid, especially a liquid, is subsequently used with a Bourdon tube, the fluid in the tube will cause a non-zero deflection according to its weight compared with air, resulting in a reading error of up to 6 per cent of full-scale deflection.

This can be avoided by calibrating the Bourdon tube with the fluid to be measured instead of with air. Alternatively, correction can be made according to the calculated weight of the fluid in the tube. Unfortunately, difficulties arise with both of these solutions if air is trapped in the tube, since this will prevent the tube being filled completely by the fluid. Then, the amount of fluid actually in the tube, and its weight, will be unknown. To avoid this problem, at least one manufacturer now provides a bleed facility in the tube which allows measurement uncertainties of less than 0.1 per cent to be achieved.

When using a *McLeod gauge*, care must be taken to ensure that the measured gas does not contain any vapour. This would be condensed during the compression process, causing a measurement error. A further recommendation is the insertion of a liquid-air cold trap between the gauge and the instrument being calibrated to prevent the passage of mercury vapour into the latter.

Mass, force and torque calibration

12.1 Review of mass measurement

Mass describes the quantity of matter which a body contains. One way in which it can be measured is to compare the gravitational force on the body with the gravitational force on another body of known mass, using a beam-balance, weigh beam, pendulum scale or electromagnetic balance, in a procedure known as 'weighing'. Alternative mass-measurement techniques are to use either a spring-balance or a load-cell. Of these, the electronic load-cell has definite advantages and is the preferred instrument nowadays in more and more industrial applications.

Beam-balance (equal-arm balance)

In the beam-balance, shown in Figure 12.1, standard masses are added to a pan on one side of a pivoted beam until the magnitude of the gravitational force on them balances the magnitude of the gravitational force on the unknown mass acting at the other end of the beam. This equilibrium position is indicated by a pointer which moves against a calibrated scale.

Instruments of this type are capable of measuring a wide span of masses. Those at the top end of the range can typically measure masses up to 1000 g whereas those at the bottom end of the range can measure masses of less than 0.01 g. Measurement resolution can be as good as 1 part in 10^7 of the full-scale reading if the instrument is designed and manufactured very carefully. The lowest measurement inaccuracy figure attainable is ±0.002 per cent.

One serious disadvantage of this type of instrument is its lack of ruggedness. Continous use and the inevitable shock loading which will occur from time

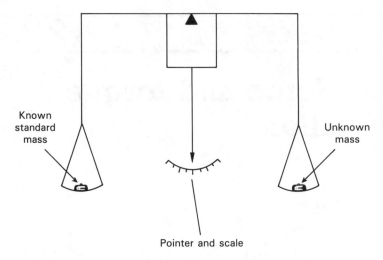

Figure 12.1 The beam balance

to time both cause damage to the knife edges, leading to problems in measurement accuracy and resolution, as discussed in Section 12.5. A further problem in industrial use is the relatively long time needed to make each measurement. For these reasons, the beam-balance is normally reserved as a calibration standard and is not used in day-to-day production environments.

Weigh beam

The weigh beam, sketched in two alternative forms in Figure 12.2, operates on similar principles to the beam-balance but is much more rugged. In the first form, standard masses are added to balance the unknown mass and fine adjustment is provided by a known mass which is moved along a notched, graduated bar until the pointer is brought to the null, balance point. The alternative form has two or more graduated bars (three bars shown in Figure 12.2). Each bar carries a different standard mass and these are moved to appropriate positions on the notched bar to balance the unknown mass. Versions of these instruments are used to measure masses up to 50 t.

Pendulum scale

The pendulum scale, sketched in Figure 12.3, is another instrument which works on the mass-balance principle. The unknown mass is put on a platform

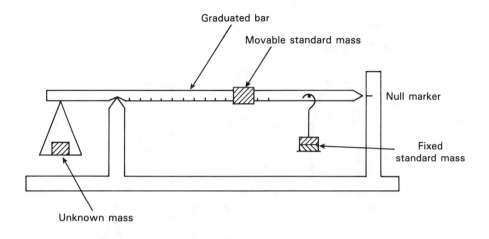

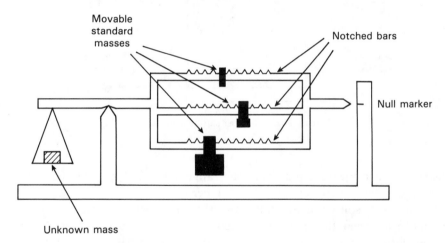

Figure 12.2 Two alternative forms of weigh beam

which is attached by steel tapes to a pair of cams. Downward motion of the platform, and hence rotation of the cams, under the influence of the gravitational force on the mass, is opposed by the gravitational force acting on two pendulum-type masses attached to the cams. The amount of rotation of the cams when the equilibrium position is reached is determined by the deflection of a pointer against a scale. The shape of the cams is such that this output deflection is linearly proportional to the applied mass.

This instrument is particularly useful in some applications because it is a relatively simple matter to replace the pointer and scale system by a rotational displacement transducer which gives an electrical output. Various versions of

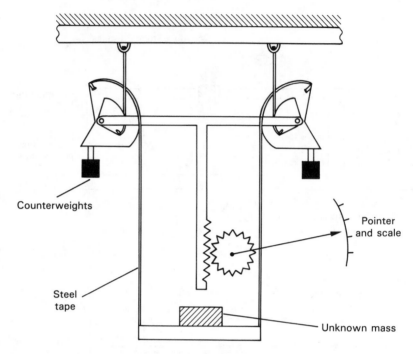

Figure 12.3 The pendulum scale

the instrument can measure masses in the range between 1 kg and 500 t, with a typical measurement inaccuracy of ±0.1 per cent.

One potential source of difficulty with the instrument is oscillation of the weigh platform when the mass is applied. Where necessary, in instruments measuring larger masses, dashpots are incorporated into the cam system to damp out such oscillations. A further possible problem can arise, mainly when measuring large masses, if the mass is not placed centrally on the platform. This can be avoided by designing a second platform to hold the mass which is hung from the first platform by knife edges. This lessens the criticality of mass placement.

Electromagnetic balance

The electromagnetic balance uses the torque developed by a current-carrying coil suspended in a permanent magnetic field to balance the unknown mass against the known gravitational force produced on a standard mass, as shown in Figure 12.4. A light source and detector system is used to determine the null-

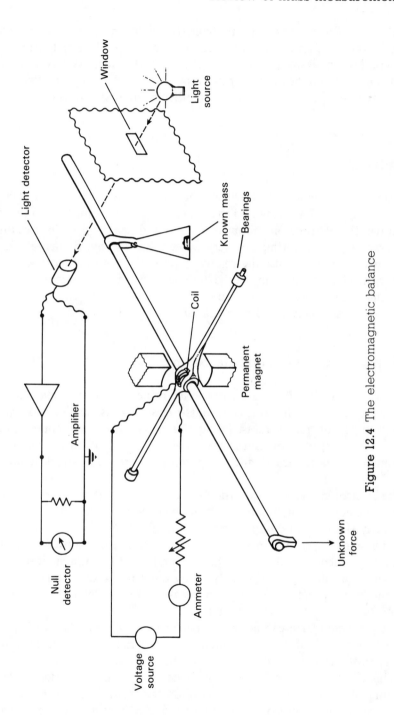

Figure 12.4 The electromagnetic balance

balance point. The voltage output from the light detector is amplified and applied to the coil, thus creating a servosystem where the deflection of the coil in equilibrium is proportional to the applied force. Its advantages over beam-balance, weigh beam and pendulum scale include its smaller size, its insensitivity to environmental changes (modifying inputs) and its electrical form of output.

Spring-balance

The spring-balance is a well-known instrument for measuring mass. The mass is hung on the end of a spring and the deflection of the spring due to the downward gravitational force on the mass is measured against a scale. Because the characteristics of the spring are very susceptible to environmental changes, measurement accuracy is usually relatively poor. If, however, compensation is made for the changes in spring characteristics, then a measurement inaccuracy of less than ±0.2 per cent is achievable. According to the design of the instrument, masses between 0.5 kg and 10 t can be measured.

Electronic load-cell (electronic balance)

Electronic load-cells, also known as electronic balances, have significant advantages over most other forms of mass-measuring instrument and so are the preferred type for most industrial applications. These advantages include relatively low cost, wide measurement range, tolerance of dusty and corrosive environments, remote measurement capability, tolerance of shock loading and ease of installation.

The electronic load-cell uses the physical principle that a force applied to an elastic element produces a measurable deflection. The elastic elements used are specially shaped and designed, some examples being shown in Figure 12.5. The design aims are to obtain a linear output relationship between the applied force and the measured deflection and to make the instrument insensitive to forces which are not applied directly along the sensing axis. Various types of displacement transducer are used to measure the deflection of the elastic elements.

One problem which can affect the performance of this class of instrument is the phenomenon of 'creep', which describes the permanent deformation which an elastic element undergoes after it has been under load for a period of time. This can lead to significant measurement errors in the form of a bias on all readings if the instrument is not re-calibrated from time to time. However, careful design and choice of materials can largely eliminate the problem.

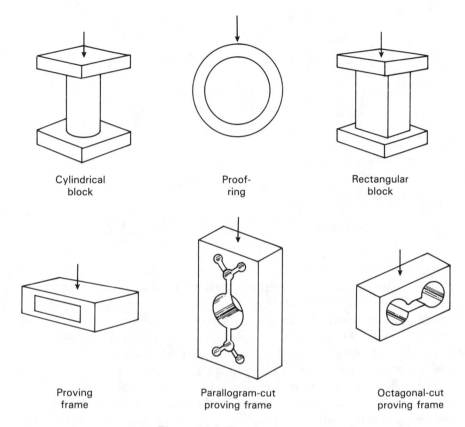

Cylindrical   Proof-   Rectangular
block   ring   block

Proving   Parallogram-cut   Octagonal-cut
frame   proving frame   proving frame

Figure 12.5 Elastic elements

The best accuracy in elastic force transducers is obtained by those instruments which use strain gauges to measure displacements of the elastic element, with an inaccuracy figure less than ±0.05 per cent of full-scale reading being obtainable. Instruments of this type are used to measure masses over a very wide range between 0 and 3×10^6 kg. The measurement capability of an individual instrument designed to measure masses at the bottom end of this range would typically be 0.1–5.0 kg, whereas instruments designed for the top of the range would have a typical measurement span of 10–3000 t.

Elastic force transducers based on differential transformers (LVDTs) to measure deflections are used to measure masses up to 25 t. Apart from having a lower maximum measuring capability, they are also inferior to strain-gauge-based instruments in terms of their ±0.2 per cent inaccuracy figure. Their major advantage is their longevity and almost total lack of maintenance requirements.

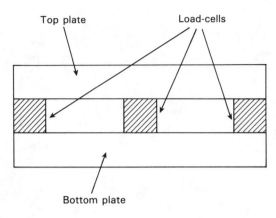

Figure 12.6 Load-cell-based electronic balance

The final type of displacement transducer used in this class of instrument is the piezo-electric device. Such instruments are used to measure masses in the range $0-10^6$ kg. Piezo-electric crystals replace the specially designed elastic member normally used in this class of instrument, allowing the device to be physically small. As discussed previously, such devices can only measure dynamically changing forces because the output reading results from an induced electrical charge whose magnitude leaks away with time. The fact that the elastic element consists of the piezo-electric crystal means that it is very difficult to design such instruments to be insensitive to forces applied at an angle to the sensing axis. Therefore, special precautions have to be taken in applying these devices. Although such instruments are relatively cheap, their lowest inaccuracy is ±1 per cent of full-scale reading, and they also have a high temperature coefficient.

All types of load-cell-based electronic balances normally contain several load-cells, as illustrated in Figure 12.6. Such arrangements are dictated by the need for stability in the mechanical construction. Usually, either three or four load-cells are used in the balance, with the output mass measurement being formed from the sum of the outputs of each cell. Where appropriate, the upper platform can be replaced by a tank for weighing liquids, powders, etc.

Pneumatic/hydraulic load-cells

Alternative forms of load-cell also exist which work on either pneumatic or hydraulic principles and convert mass measurement into a pressure-measurement problem. A pneumatic load-cell is shown schematically in Figure 12.7. Application of a mass to the cell causes deflection of a diaphragm acting

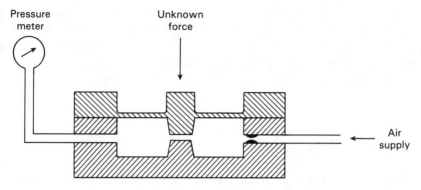

Figure 12.7 Pneumatic load-cell

as a variable restriction in a nozzle–flapper mechanism. The output pressure measured in the cell is approximately proportional to the magnitude of the gravitational force on the applied mass. The instrument requires a flow of air at its input of around 0.25 m³/h at a pressure of 4 bar. Standard cells are available to measure a wide range of masses. For measuring small masses, instruments are available with a full-scale reading of 25 kg, while at the top of the range, instruments with a full scale reading of 25 t are obtainable. Inaccuracy is typically ±0.5 per cent of full scale in pneumatic load-cells.

The alternative, hydraulic load-cell is shown in Figure 12.8. In this, the gravitational force due to the unknown mass is applied, via a diaphragm, to oil contained within an enclosed chamber. The corresponding increase in oil pressure is measured by a suitable pressure transducer. These instruments are designed for measuring much larger masses than pneumatic cells, with a load capacity of 500 t being common. Special units can be obtained to measure

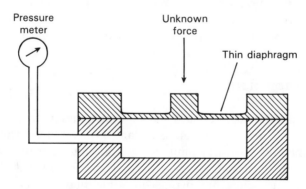

Figure 12.8 Hydraulic load-cell

masses as large as 50 000 t. Besides their much greater measuring range, hydraulic load-cells are much more accurate than pneumatic cells, with an inaccuracy figure of ±0.05 per cent of full scale being typical, although to obtain this level of accuracy, correction for the local value of g (acceleration due to gravity) is necessary. A measurement resolution of ±0.02 per cent is attainable.

Intelligent load-cells

Intelligent load-cells are formed by adding a microprocessor to a standard cell. This brings no improvement in accuracy because the load-cell is already a very accurate device. What it does produce is an intelligent weighing system which can compute total cost from the measured weight, using stored cost per unit weight information, and provide an output in the form of a digital display. Cost per weight figures can be pre-stored for a large number of substances, making the instrument very flexible in its application.

12.2 Review of force measurement

If a force of magnitude, F, is applied to a body of mass, M, the body will accelerate at a rate, A, according to the equation

$$F = M \cdot A$$

The standard unit of force is the Newton, this being the force which will produce an acceleration of 1 m/s^2 in the direction of the force when it is applied to a mass of 1 kg. One way of measuring an unknown force is therefore to measure the acceleration when it is applied to a body of known mass.

An alternative technique is to measure the variation in the resonant frequency of a vibrating wire as it is tensioned by an applied force.

Use of accelerometers

The technique of applying a force to a known mass and measuring the acceleration produced can be carried out using any type of accelerometer. Unfortunately, the method is of very limited practical value because, in most cases, forces are not free entities but are part of a system (from which they cannot be decoupled) in which they are acting on some body which is not free to accelerate. However, the technique can be of use in measuring some transient forces, and also for calibrating the forces produced by thrust motors in space vehicles.

Vibrating wire sensor

This instrument, illustrated in Figure 12.9, consists of a wire which is kept vibrating at its resonant frequency by a variable-frequency oscillator. The resonant frequency of a wire under tension is given by

$$f = \frac{0.5}{L} \left[\frac{M}{T} \right]^{1/2}$$

where M is the mass per unit length of the wire, L is the length of the wire, and T is the tension due to the applied force, F.

Thus, measurement of the output frequency of the oscillator allows the force applied to the wire to be calculated.

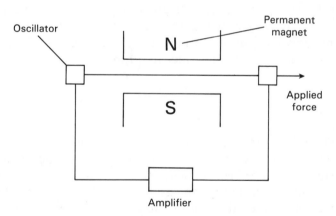

Figure 12.9 The vibrating wire sensor

12.3 Review of torque measurement

Measurement of applied torques is of fundamental importance in all rotating bodies to ensure that the design of the rotating element is adequate to prevent failure under shear stresses. Torque measurement is also a necessary part of measuring the power transmitted by rotating shafts.

Four methods of measuring torque exist, as follows:

(a) measuring the reaction force in cradled shaft bearings;
(b) the 'Prony brake' method;
(c) measuring the strain produced in a rotating body due to an applied torque; and
(d) optical torque measurement.

The first three of the above methods are traditional ones and the fourth is a new technique made possible by recent developments in electronics and fibre-optic technology.

Reaction forces in shaft bearings

Any system involving torque transmission through a shaft contains both a power source and a power absorber where the power is dissipated. The magnitude of the transmitted torque can be measured by cradling either the power source or the power absorber end of the shaft in bearings and then measuring the reaction force, F, and the arm length L, as shown in Figure 12.10(a) and (b). The torque is then calculated as the simple product, $F \cdot L$. Pendulum scales are very commonly used for measuring the reaction force. Inherent errors in the method are bearing friction and windage torques.

Prony brake

The principle of the Prony brake is illustrated in Figure 12.11. It is used to measure the torque in a rotating shaft and consists of a rope wound round the shaft, one end of the rope being attached to a spring-balance and the other end carrying a load in the form of a standard mass, m. If the measured force in the spring balance is F_s, then the effective force, F_e, exerted by the rope on the shaft is given by

$$F_e = mg - F_s$$

If the radius of the shaft is R_s and that of the rope is R_r, then the effective radius, R_e, of the rope and drum with respect to the axis of rotation of the shaft is given by

$$R_e = R_s + R_r$$

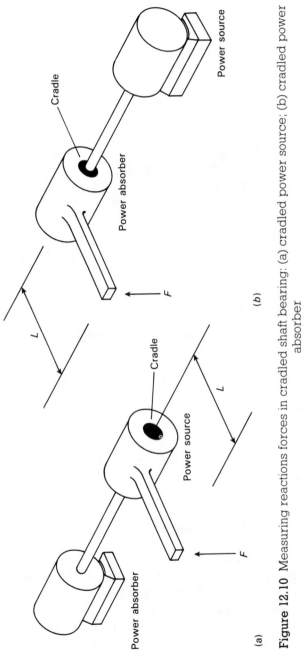

Figure 12.10 Measuring reactions forces in cradled shaft bearing: (a) cradled power source; (b) cradled power absorber

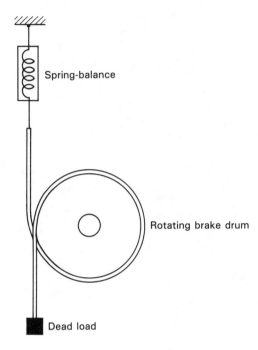

Figure 12.11 Prony brake

The torque in the shaft, T, can then be calculated as

$$T = F_e \cdot R_e$$

While this is a well-known method of measuring shaft torque, a lot of heat is generated because of friction between the rope and shaft, and water cooling is usually necessary.

Measurement of induced strain

Measuring the strain induced in a shaft due to an applied torque has been the most common method used for torque measurement in recent years. It is a very attractive method because it does not disturb the measured system by introducing friction torques in the same way as the last two methods described. The method involves bonding four strain gauges onto the shaft, connected in a dc bridge circuit, as shown in Figure 12.12. The output from the bridge circuit is a function of the strain in the shaft and hence of the torque applied. It is very

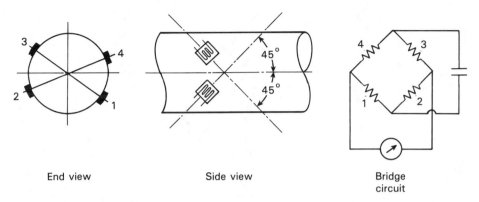

| End view | Side view | Bridge circuit |

Figure 12.12 Position of torque-measuring strain gauges on shaft

important that the positioning of the strain gauges on the shaft is precise, and the difficulty in achieving this makes the instrument relatively expensive.

The technique is ideal for measuring the stalled torque in a shaft before rotation commences. However, a problem is encountered in the case of rotating shafts because a suitable method then has to be found for making the electrical connections to the strain gauges. One solution to this problem found in many commercial instruments is to use a system of slip rings and brushes, although this increases the cost of the instrument still further.

Optical torque measurement

Optical techniques for torque measurement have become available recently with the development of laser diodes and fibre-optic light transmission systems. One such system is shown in Figure 12.13. Two black and white striped wheels are mounted at either end of the rotating shaft and are in alignment when no torque is applied to the shaft. Light from a laser diode light source is directed by a pair of optic-fibre cables onto the wheels. The rotation of the wheels causes pulses of reflected light and these are transmitted back to a receiver by a second pair of fibre-optic cables. Under zero-torque conditions, the two pulse-trains of reflected light are in phase with each other.

If torque is then applied to the shaft, the reflected light is modulated. Measurement by the receiver of the phase difference between the reflected pulse-trains therefore allows the magnitude of torque in the shaft to be calculated. The cost of such instruments is relatively low, and an additional advantage in many applications is their small physical size.

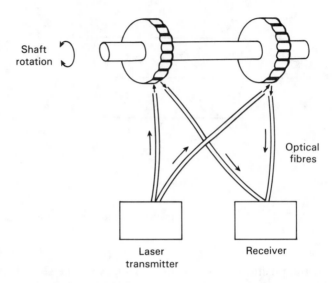

Figure 12.13 Optical torque measurement

12.4 Introduction to calibration

A complication which arises in the calibration of mass, force and torque measuring instruments is the variability in the value of g (the acceleration due to gravity). Apart from instruments such as the beam-balance and pendulum scale which directly compare two masses, all other instruments have an output reading which depends on the value of g.

The value of g is given by Helmert's formula:

$$g = 980.6 - 2.6 \cos \Phi - 0.000\ 309\ h$$

where Φ is the latitude and h is the altitude in metres. It can be seen from this formula that g varies with both latitude and altitude. At the equator ($\cos \Phi = 0°$) $g = 978.0$, whereas at the poles ($\cos \Phi = 90°$) $g = 983.2$. In Britain, a working value of 980.7 is normally used for g, and very little error can normally be expected when using this value. Where necessary, the exact value of g can be established by measuring the period and length of a pendulum.

A further difficulty which arises in calibrating instruments in this category is the upwards force generated by the air medium in which the instruments are tested and used. According to Archimedes' principle, when a body is immersed in a fluid (air in this case), there is an upwards force proportional to the volume of fluid displaced. Even in pure mass-balance instruments, an error is introduced because of this unless both the body of unknown mass and the standard mass have the same density. This error is given by

$$\xi = \frac{SG_a}{SG_u} - \frac{SG_a}{SG_m}$$

where SG_a is the specific gravity of air, SG_u is the specific gravity of the substance being measured and SG_m is the specific gravity of the standard mass.

Fortunately, the maximum error due to this (which occurs when weighing low-density liquids such as petrol) will not exceed 0.2 per cent. Therefore, in most circumstances, the error due to air buoyancy can be neglected. However, for calibrations at the top of the calibration tree, where the highest levels of accuracy are demanded, either correction must be made for this factor or it must be avoided by carrying out the calibration in vacuum conditions.

12.5 Calibration instruments and equipment

The primary requirement in mass calibration is the maintenance of a set of standard masses against which the mass to be calibrated can be compared.[1] Beam-balances, weigh beams, pendulum scales, electromagnetic balances and load-cells are all used in various circumstances to effect this comparison.

Beam-balance

The instrument used most commonly for calibrating masses is the beam-balance. This is used for calibrations in the range between 10 mg and 1 kg. The measurement resolution and accuracy of such instruments depends on the quality and sharpness of the knife edge that the pivot is formed from. For high measurement resolution, friction at the pivot must be as close to zero as possible and hence a very sharp and clean knife-edge pivot is demanded. The two halves of the beam on either side of the pivot are normally of equal length and are measured from the knife edge. Any bluntness, dirt or corrosion in the pivot can cause these two lengths to become unequal, causing consequent measurement errors. Similar comments apply to the knife edges on the beam that the two pans are hung from. It is also important that all knife edges are parallel, otherwise displacement of the point of application of the force over the line of the knife edge can cause further measurement errors. This last form of error also occurs if the mass is not placed centrally on the pan.

Great care is therefore required in the use of such an instrument but, provided that it is kept in good condition, particularly with regard to keeping the knife edges sharp and clean, high measurement accuracy is achievable.

Weigh beam

The weigh beam, when manufactured and maintained at a sufficient standard of accuracy, can be used as a standard for masses up to 50 t.

Pendulum scale

In the range above 1 kg, up to 500 t, versions of the pendulum scale can be used for mass calibration. Again, such calibration instruments must be kept in good condition, with special attention to the cleanliness and lubrication of moving parts.

Electromagnetic balance

Various forms of electromagnetic balance exist as alternatives to the last three instruments. Their optical system magnifies motion around the null point, leading to greater accuracy. Consequently, this type of instrument is preferred for higher-level calibrations. The actual degree of accuracy achievable depends on the magnitude of the mass being measured. In the range between 100 g and 10 kg, an inaccuracy of ± 0.0001 per cent is achievable. Above and below this range, the inaccuracy is worse, increasing to ± 0.002 per cent measuring 5 t and ± 0.03 per cent measuring 10 mg.

Proof-ring-based load-cell

In the range between 150 kg and 2000 t, a proof-ring-based load-cell is used for calibration in which the displacement of the proof ring is measured by either an LVDT or a micrometer. As the relationship between the applied mass–force and the displacement is not a straight-line one, a force–deflection graph has to be used to interpret the output. Minimum measurement inaccuracy is ± 0.1 per cent.

Shaft with known torque applied

Torque-measuring instruments are calibrated by applying them to a shaft to which a known torque is applied.

12.6 Calibration frequency

The primary requirement in mass calibration is a set of standard masses for comparison. Provided that these do not suffer any obvious damage, annual calibration checks are quite adequate.

The required re-calibration frequency for instruments used to effect comparison of standard masses is somewhat more difficult to define as it depends on the conditions of use. Therefore, in the case of the beam-balance, weigh beam, pendulum scale and electromagnetic balance, practical measurement of the rate of degradation in measurement accuracy is necessary in order to determine a suitable re-calibration frequency.

Both elastic-element-based load-cells and (especially) spring-balances suffer from elastic deformation of the active element. Again, practical measurement of the rate of accuracy degradation is required but, as a general rule, re-calibration of such instruments will have to be carried out much more frequently than for mass-comparison instruments.

13

Dimension calibration

13.1 Review of dimension-measuring instruments

Dimension measurement includes the measurement of the length, height, depth (of holes and slots) and angles of components. For many such measurements, it is necessary to have a reference flat plane on which components being measured are located. This is provided by a surface plate or table as described below.

The range of instruments available for measuring length includes the steel tape (to 30 m), ultrasonic rule (to 10 m), steel rule (to 1 m), standard calipers (to 600 mm), vernier calipers (to 200 mm) and micrometer (to 50 mm). Gauge blocks and length bars, although primarily intended for calibration duties, are also used for the measurement of larger lengths when very high accuracy is required. Height and depth are measured by the height gauge, depth gauge or dial gauge. Angles are normally measured by either an angle protractor, a bevel protractor or a spirit level. Further information on dimension measurement can be found in references 1 and 2.

Surface planes and tables

A flat and level reference plane is often an essential component in dimension measurement, especially where the line of measurement and the dimension being measured are not coincident. Such reference planes are available in a range of standard sizes, the smallest having nominal dimensions of 100 mm × 160 mm and the largest 1600 mm × 2500 mm. The smaller sizes exist as a surface plate resting on a supporting table, whereas the larger sizes take the form of free standing tables, as shown in Figure 13.1. Larger sizes have a

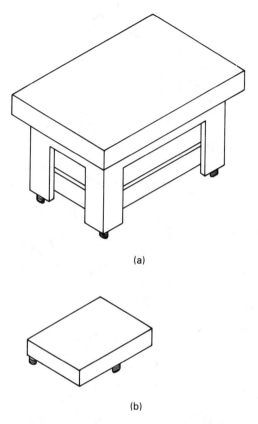

(a)

(b)

Figure 13.1: (a) surface table; (b) surface plate

projection at the edge to facilitate the clamping of components. They are normally used in conjunction with box cubes and vee blocks (see Figure 13.2) which locate components in a fixed position. The plate or table normally has three feet,* with provision for levelling to make the surface exactly horizontal. In modern tables, granite has tended to supersede iron as the preferred material for the plate, although iron plates are still available. Granite is ideal for this purpose as it does not corrode, is dimensionally very stable and does not form burrs when damaged. Iron plates, on the other hand, are prone to rusting and susceptible to damage; this results in burrs on the surface which interfere with measurement procedures.

Both forms of plate are available in four standard grades, from grade 0 to grade 3. Grade 0 is the best and is used for calibration purposes. Grades 1–3

---

*plates with dimensions of 1 m × 1.6 m and larger often have more than three feet.

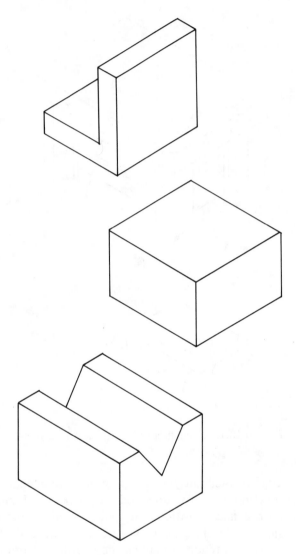

Figure 13.2 Box cubes and vee blocks

are recommended for inspection, marking out and lower grade marking out respectively. Flatness is defined in terms of the distance between two parallel planes which just contain all points in the surface. Standards of flatness, as defined by *BS 817*,[3] vary according to the size of the plate or table. For a 2 m × 1 m table, the maximum permitted deviations from flatness for grades 0–3 are 8.5 μm, 17 μm, 34 μm, 68 μm respectively.

Steel rule

The steel rule is undoubtedly the simplest instrument available for measuring length. Rules are manufactured according to quality standards (see reference 4) which define the ruling accuracy, squareness of the end and edge straightness. The best rules have rulings at 0.05 mm intervals and a measurement resolution of 0.02 mm. When used by placing the rule against an object, the measurement accuracy is much dependent upon the skill of the human measurer and, at best, the inaccuracy is likely to be at least ±0.5 per cent.

Steel tape

The retractable steel tape is another well-known instrument. The end of the tape is usually provided with a flat hook which is loosely fitted so as to allow for automatic compensation of the hook thickness when the rule is used for both internal and external measurements. Again, measurement accuracy is governed by human skill but, with care, the measurement inaccuracy achieved can be as low as ±0.01 per cent. Formal specifications for measuring tapes are defined in references 5 and 6.

Ultrasonic rule

The ultrasonic rule is a recent addition to the measurement scene. It consists of an ultrasonic energy source, an ultrasonic energy detector and battery-powered, electronic circuitry housed within a hand-held box, as shown in Figure 13.3. Both source and detector often consist of the same type of piezo-electric crystal

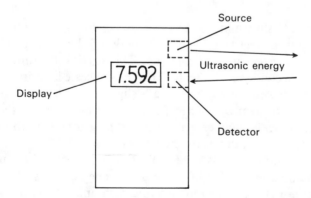

Figure 13.3 Ultrasonic rule

excited at a typical frequency of 40 kHz. Energy travels from the source to a target object and is then reflected back into the detector. The time of flight of this energy is measured and this is converted into a distance reading by the enclosed electronics. Maximum measurement inaccuracy of ±1 per cent of the full-scale reading is claimed. This is only a modest level of accuracy which is of little use for most engineering measurements. However, it is sufficient for such purposes as room measurements taken by estate agents prior to producing sales literature, where the ease and speed of making measurements is of great value.

A fundamental problem in the use of ultrasonic energy of this type is the limited measurement resolution (7 mm) imposed by the 7 mm wavelength of sound at this frequency. Further problems are caused by the variation in the speed of sound with humidity (variations of ±0.5 per cent possible) and the temperature-induced variation of 0.2 per cent/°C. Therefore, the conditions of use must be carefully controlled if the claimed accuracy figure is to be met.

Standard caliper

Figure 13.4 shows two types of standard caliper. These are used to transfer the measured dimension from the workpiece to a steel rule. This avoids the necessity to align the end of the rule exactly with the edge of the workpiece and reduces the measurement inaccuracy by a factor of two. In the basic caliper, careless use can allow the setting of the caliper to be changed during transfer from the workpiece to the rule. Hence, the spring-loaded type, which prevents this happening, is preferable.

Vernier caliper

Vernier calipers,[7,8] shown in Figure 13.5, are effectively a combination of standard calipers and a steel rule. The main body of the instrument includes the main scale with a fixed anvil at one end. This carries a sliding anvil which is provided with a second, vernier scale. This second scale is shorter than the main scale and is divided into units which are slightly smaller than the main scale units but related to them by a fixed factor. Determination of the point where the two scales coincide enables very accurate measurements to be made, with typical inaccuracy levels down to ±0.01 per cent.

Figure 13.6 shows details of a typical combination of main and Vernier scales. The main scale is ruled in 1 mm units. The vernier scale is 49 mm long and divided into fifty units, thereby making each unit 0.02 mm smaller than the main scale units. Each group of five units on the vernier scale thus differs from

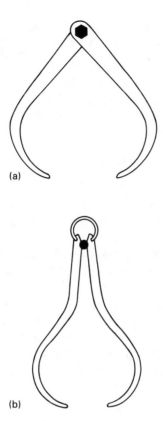

Figure 13.4: (a) standard caliper; (b) spring-loaded caliper

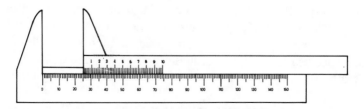

Figure 13.5 Vernier caliper

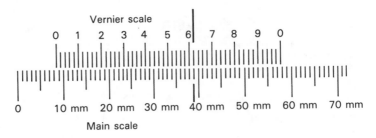

Figure 13.6 Details of Vernier caliper scale

the main scale by 0.1 mm and the numbers marked on the scale thus refer to these larger units of 0.1 mm. In the particular position shown in the Figure 13.6, the zero on the vernier scale indicates a measurement between 8 and 9 mm. Both scales coincide at a position of 6.2 (large units). This defines the interval between 8 and 9 mm to be $6.2 \times 0.1 = 0.62$ mm, i.e. the measurement is 8.62 mm.

Intelligent digital calipers are now available which give a measurement resolution of 0.01 mm and an accuracy of ± 0.03 mm. These have automatic compensation for wear, and hence calibration checks have to be very infrequent. In some such instruments, the digital display can be directly interfaced to an external computer monitoring system.

Micrometers

In the standard micrometer, shown in Figure 13.7(a), measurement is made between two anvils, one fixed and one which is moved along by the rotation of an accurately machined screw thread. One complete rotation of the screw typically moves the anvil by a distance of 0.5 mm. Such movements of the anvil are measured using a scale marked with divisions every 0.5 mm along the barrel of the instrument. A scale marked with fifty divisions is etched around the circumference of the spindle holder; each division therefore corresponds to an axial movement of 0.01 mm. Assuming that the user is able to judge the position of the spindle on this circumferencial scale against the datum mark to within one-fifth of a division, a measurement resolution of 0.002 mm is possible. Such instruments are manufactured to high standards of accuracy, and typical specifications might be: flatness of anvils, ± 0.001 mm; parallelism of anvils, ± 0.003 mm; screw traverse, ± 0.003 mm; alignment error, 0.05 mm max.[9,10]

The most common measurement ranges are either 0–25 mm or 25–50 mm, with inaccuracy levels down to ± 0.003 per cent. However, a whole family of

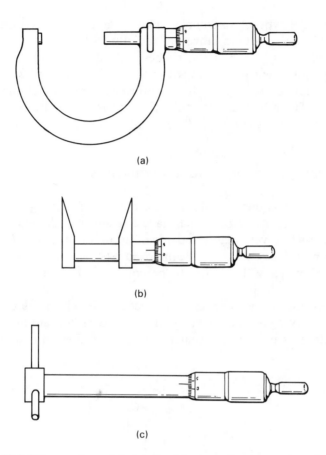

(a)

(b)

(c)

Figure 13.7 Micrometers: (a) standard (external) micrometer; (b) internal micrometer; (c) bore micrometer

micrometers is available, where each has a measurement span of 25 mm, but with the minimum distance measured varying from 0 mm up to 575 mm. Thus, the last instrument in this family measures the range from 575 to 600 mm. Some manufacturers also provide micrometers with two or more interchangeable anvils, which extends the span measurable with one instrument to between 50 mm and 100 mm according to the number of anvils supplied. Therefore, an instrument with four anvils might, for instance, measure the range from 300 mm to 400 mm by making appropriate changes to the anvils.

An alternative form of micrometer (see Figure 13.7(b)) is able to measure internal dimensions such as hole diameters.[11] In the case of measuring holes, micrometers are inaccurate if there is any ovality in the hole, unless the diameter is measured at several points. An alternative solution to this problem is to use a special type of instrument, known as a bore micrometer (Figure 13.7(c)). In this, three probes move out radially from the body of the instrument as the spindle is turned. These probes make contact with the sides of the hole at three equidistant points, thus averaging out any ovality.

Intelligent micrometers in the form of the electronic digital micrometer are now available. These have a self-calibration capability and a digital readout, with a measurement resolution of 0.001 mm (1 micron).

Gauge blocks (slip gauges) and length bars

Gauge blocks, also known as slip gauges, consist of rectangular blocks (see Figure 13.8) of hardened steel which have flat and parallel end faces. These faces are machined to very high standards of accuracy in terms of their surface finish and flatness. They are available in boxed sets containing a range of block sizes, which allows any dimension to be constructed by joining together an appropriate number of blocks.

Blocks are joined by 'wringing', a procedure in which the two end faces are rotated slowly against each other. This removes the air film and allows adhesion to develop by inter-molecular attraction. Adhesion is so good in fact

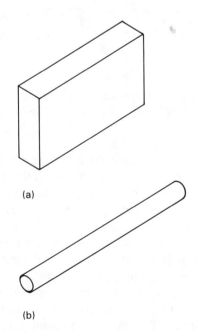

(a)

(b)

Figure 13.8: (a) gauge block; (b) length bar

that, if groups of blocks were not separated within a few hours, the molecular −
diffusion process would continue to the point where the blocks would be permanently welded together. The typical inter-block gap resulting from wringing has
been measured as 0.001 μm, which is effectively zero. Thus, any number of
blocks can be joined without creating any significant measurement error.

Gauge blocks are available in five standards of accuracy: calibration, 00,
0, 1 and 2. Grades 1 and 2 are used for normal production inspection measurements. The length, end-face-flatness and end-face-parallelism tolerences
allowed for gauge blocks are defined by *BS 4311.*[*][12]

For a 100 mm long block, the tolerances are:

	Length	Flatness	Parallelism
Grade 1:	+0.6 μm −0.3 μm	±0.15 μm	±0.25 μm
Grade 2:	+1.4 μm −1.0 μm	±0.25 μm	± 0.35μm

The larger positive tolerance for grade 1 and 2 blocks is provided to allow for
the relatively high level of wear which occurs immediately after a new block is
put into use. It should be noted that the length tolerances of blocks wrung
together are cumulative. Thus the length tolerance of five grade 2 blocks wrung
together is (+7 μm, −5 μm).

A typical set of gauge blocks allows any dimension between 3.0 mm and
200 mm to be built up in steps of 0.0005 mm. By careful choice of the block
combination, it is usually possible to construct any dimension using not more
than five blocks. Consider the M112/1 set below which consists of the following
112 blocks:

Size or series	Increment	No. of pieces
1.0005	−	1
1.001−1.009	0.001	9
1.01−1.49	0.01	49
0.5−24.5	0.5	49
25−100	25	4
		112

[*] ISO 3650 defines similar, though not identical, standards.[13]

The procedure for choosing appropriate blocks to make up a particular length is to consider each decimal point in turn, as follows:

1. Choose the thinnest block to satisfy the last decimal place.
2. Choose the next block to satisfy the next-to-last decimal place.
3. Continue in this fashion until the full length required is constructed.

For example to construct a length of 89.4365 mm, the following five blocks would be used:

```
 1.0005
 1.0060
 1.4300
11.0000
75.0000
89.4365
```

It is fairly common practice with blocks of grades 0, 1 and 2 to include an extra pair of 2 mm-thick blocks in the set which are made from wear-resisting tungsten carbide. These are marked with a letter P and are designed to protect the other blocks from wear during use. Where such protector blocks are used, due allowance has to be made for their thickness (4 mm) in calculating the sizes of block needed to make up the required length.

A precaution to be followed when using gauge blocks is to avoid handling them more than is necessary. The length of a bar which was 100 mm long at 20 °C would increase to 100.02 mm at 37 °C (body temperature). Hence, after wringing bars together, they should be left to stabilize back to the ambient room temperature before use. This wait might need to be several hours duration if the blocks have been handled to any significant extent.

Where a greater dimension than 200 mm is required, gauge blocks are used in conjunction with *length bars*; these consist of straight, hardened, high-quality steel bars of a uniform 22 mm diameter and in a range of lengths between 100 mm and 1200 mm. They are available in four standards of accuracy, as defined by *BS 5317*;[14] reference, calibration, grade 1 and grade 2. Reference and calibration grades have accurately flat end faces, which allows a number of bars to be wrung together to obtain the required standard length. Bars of grades 1 and 2 have threaded ends which allows them to be screwed together. Grade 2 bars are used for general measurement duties, with grade 1 bars being reserved for inspection duties. By combining length bars with gauge blocks, any dimension up to about 2 m can be set up with a resolution of 0.0005 mm.

The length tolerences of a 200 mm length bar are given by

Grade 1: $+1.4 \atop -0.6$ $\times (0.2 + 0.004L)$ μm, i.e. $+1.4$ μm or -0.6 μm

Grade 2: $+1.4 \atop -0.6$ $\times (0.4 + 0.006L)$ μm, i.e. $+2.2$ μm or -1.0 μm

where L is the length of the bar in mm. As for gauge blocks, the larger positive tolerance for grades 1 and 2 is provided to allow for the high level of wear occurring immediately after new bars are put into use. These length tolerances are only valid if the bar is horizontal and at 20 °C. Horizontal alignment is obtained by mounting the bars at their 'airy' points. The airy points are at a distance of $0.2117L$ from the ends of the bar and are marked by circumferential lines.

Height and depth gauges

The height gauge,[15] shown in Figure 13.9(a), effectively consists of a vernier caliper mounted on a flat base. Measurement inaccuracy levels down to ±0.015 per cent are possible. The depth gauge (Figure 13.9(b)) is a further variation on the standard vernier caliper principle which has the same measurement accuracy capabilities as the height gauge.

In practice, certain difficulties can arise in the use of these instruments where either the base of the instrument is not properly located on the measuring table or the point of contact between the moving anvil and the workpiece is uncertain. In such cases, a dial gauge, which has a clearly defined point of contact with the measured object, is used in conjunction with the height or depth gauge to avoid these possible sources of error.

These instruments can also be obtained in intelligent versions which give a digital display and have self-calibration capabilities.

Dial gauge

The dial gauge,[16,17] shown in Figure 13.10, consists of a spring-loaded probe which drives a pointer around a circular scale via rack and pinion gearing. Typical measurement resolution is 0.01 mm. When used to measure the height of objects, it is clamped in a retort stand and a measurement is taken of the height of the unknown component. Then it is put in contact with a height gauge (Figure 13.11) which is adjusted until the reading on the dial gauge is the same.

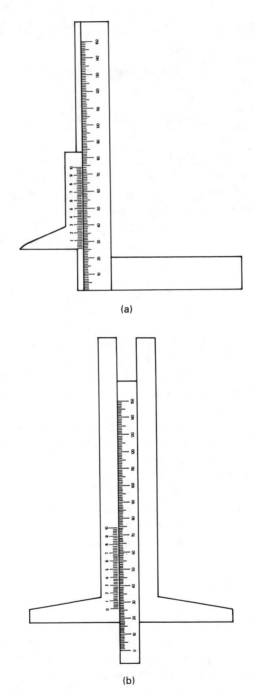

(a)

(b)

Figure 13.9: (a) height gauge; (b) depth gauge

Figure 13.10 Dial gauge

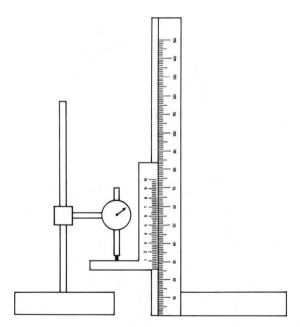

Figure 13.11 Use of dial gauge in conjunction with height gauge

At this stage, the height gauge is set to the height of the object. The dial gauge is also used in conjunction with the depth gauge in an identical manner. (Gauge blocks can be used instead of height–depth gauges in such measurement procedures if greater accuracy is required.)

Protractors

The simplest form of protractor is the angle type shown in Figure 13.12(a). This consists of two straight edges, one of which is able to rotate with respect to the

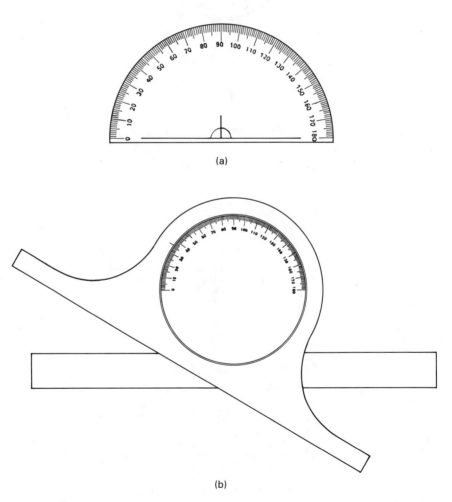

Figure 13.12 Angle measurement: (a) angle protractor; (b) bevel protractor

other. A circular scale attached to one of the straight edges rotates inside a fixed circular housing attached to the other straight edge. The relative angle between the two straight edges in contact with the component being measured is determined by the position of the moving scale with respect to a reference mark on the fixed housing. With this type of instrument, measurement inaccuracy is at least ±1 per cent.

A more accurate instrument is the bevel protractor shown in Figure 13.12(b).[18] This is of the same form as the angle protractor, but it has a vernier scale on the fixed housing. This allows the inaccuracy level to be reduced to ±10 minutes of arc.

Spirit level

The spirit level shown in Figure 13.13 is an alternative angle-measuring instrument. It consists of a standard spirit level attached to a rotatable circular scale which is mounted inside an accurately machined square frame. When placed on the sloping surfaces of components, rotation of the scale to centralize the bubble in the spirit level allows the angle of slope to be measured. Again,

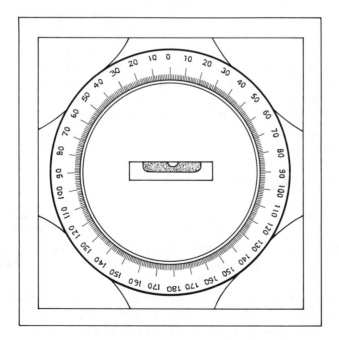

Figure 13.13 Angle-measuring spirit level

measuring inaccuracies down to ±10 minutes of arc are possible if a vernier scale is incorporated in the instrument.[19]

Electronic spirit level

Electronic spirit levels contain a pendulum whose position is sensed electrically. With such instruments, measurement resolutions as good as 0.2 seconds of arc are possible.

13.2 'User' calibration

When dimension measurements are being made as part of a quality control system, checks on the way that the human operator is using equipment are just as important as calibration checks on the instruments themselves. Such 'calibration' of the equipment user is very important because there are so many ways in which human-induced measurement errors can be introduced. The scope for human-induced errors is far greater than that which exists when measuring most other physical quantities.

The golden rule in dimension measurement is that the line of measurement and the line of the dimension being measured should be coincident. In the case of steel rules and tapes, the greatest potential source of user-induced error is failure to position the rule squarely across the dimension being measured. Parallax error is also possible if the user does not position the rule and read it from directly above. Calipers and micrometers are less susceptible to these types of error but it still remains sensible to carry out periodic checks on the way in which these instruments are being used, verifying in particular that measurements are being made squarely.

13.3 Calibration equipment

The equipment needed for dimension calibration are a set of gauge blocks, a set of end bars, a reference plane (horizontal table) and a sine bar for calibrating angle-measuring instruments. Optically flat glass plates are also needed for some procedures which involve testing the flatness and parallelism of metal faces.

Calibration is performed by comparing measurements made on the reference plane with the unknown instrument and the standard. Reference standards for length consist of either gauge blocks alone or gauge blocks and end bars together for testing larger dimensions. A high-magnification comparator is required to make this comparison. At the very highest levels of calibration of gauge blocks themselves, interferometric methods are used as a primary reference standard. Such calibration services are normally provided by national standards organizations (e.g. National Physical Laboratory in the United Kingdom) or specialized calibration companies.

Setting gauges also deserve mention in respect of their use for initial setting up of new internal micrometers and subsequent calibration checks of these instruments.

Reference plane

A reference plane is an essential component in dimension calibration. Surface plates and tables of flatness grade 0 or 1 are used according to the level of calibration. They are constructed either of graphite or iron, as discussed in Section 13.1.

Proper procedures for care and use of reference planes used for calibration duties are defined in *BS 817*. These require the surface to be wiped clean of dust at frequent intervals, to be used carefully to avoid damage and to be protected with a purpose-designed cover when not in use. Use in conditions of circulating air under constant temperature control, avoiding direct sunlight and draughts, is recommended.

The reference plane itself must be subjected to periodic calibration checks to test its flatness and horizontal alignment. The variation gauge (Figure 13.14) is available for measuring flatness. This has three fixed feet in contact with the flat surface and a floating foot whose position is indicated on a dial gauge.

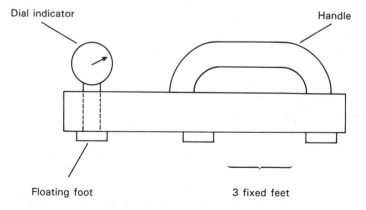

Dial indicator

Handle

Floating foot

3 fixed feet

Figure 13.14 Variation gauge

However, flatness measurement is normally entrusted to specialist companies who maintain equipment to carry out such checks which includes computational power to simplify the analysis of surface measurements.

Gauge blocks (slip gauges) and length bars

Gauge blocks are available in five standards of accuracy: calibration, 00, 0, 1 and 2. The permitted tolerences on length, flatness and end-face-parallelism for the top three grades of bar, as defined by *BS 4311* are as follows for a 100 mm block:[12]

	Length	*Flatness*	*Parallelism*
Calibration grade:	±0.5 μm	±0.05 μm	±0.1 μm
Grade 00:	±0.15 μm	±0.05 μm	±0.1 μm
Grade 0:	±0.25 μm	±0.1 μm	±0.15 μm

The standard procedure in calibrating gauge blocks themselves is to compare them against gauge blocks of the next highest accuracy standard. Thus grade 2 is calibrated against grade 1, grade 1 against grade 0, etc. In performing this comparison, it is very important to check the flatness and degree of parallelism of the opposing end faces of each block, as these latter two parameters are just as important as length in determining the possible error in total length when a number of blocks are wrung together. Above grade 0, interferometric methods provide better reference standards for length than 'calibration standard' gauge blocks, with inaccuracy levels down to ±0.03 μm being possible. Because of this, 'calibration standard' blocks are only provided for checking the flatness and parallism of gauge 0 blocks, with length being checked by interferometric methods. However, for situations where such alternative reference length standards are not available, special grade 00 blocks are available. These have the same standards of flatness and parallelism as 'calibration grade' blocks but they have a tighter length tolerence. The calibration chains for gauge blocks are summarized in Figure 13.15.

Length bars are available in four standards of accuracy:[14] reference, calibration, grade 1 and grade 2. As for gauge blocks, the standard calibration procedure is to compare them against length bars of the next highest accuracy standard. Thus, grade 2 bars are calibrated against grade 1 bars, grade 1 against 'calibration' grade, etc. Reference and calibration grades are calibrated themselves by interferometric methods, where inaccuracy levels down to ±0.5 μm for a 1 m bar are achievable. Reference and calibration grades have accurately

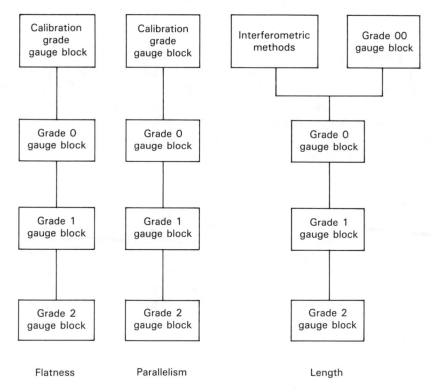

Figure 13.15 Calibration chains for gauge blocks

flat end faces, allowing a number of bars to be wrung together to obtain the required standard length. By combining length bars with gauge blocks, any dimension up to about 2 m can be set up with a resolution of 0.0005 mm.

The length tolerences of a 200 mm length bar are given by

Grade 1: $^{+1.4}_{-0.6} \times (0.2 + 0.004L)$ μm, i.e. $+1.4$ μm or -0.6 μm.

Reference grade: $(0.05 \pm 0.0015L)$ μm, i.e. ± 0.35 μm.

Calibration grade: $(0.1 \pm 0.003L)$ μm, i.e. ± 0.65 μm.

These length tolerences are only valid if the bar is horizontal and at 20 °C. Horizontal alignment is obtained by mounting the bars at their 'airy' points. The airy points are at a distance of $0.2117L$ from the ends of the bar and are marked by circumferential lines. Reference and calibration grades of bar must only be used in a standards laboratory where the temperature is accurately controlled at 20 °C.[20]

Setting gauges

Setting gauges are used for setting up new internal micrometers and re-calibrating those in use. For testing instruments measuring up to 50 mm, the setting gauge consists of a 25 mm diameter steel disc (with ±0.001 mm tolerance on its diameter). For testing larger range instruments, the setting gauge consists of either a flat-ended or a spherical-ended steel rod, with a length tolerance which varies from ±0.002 mm on a 125 mm long gauge to ±0.006 mm on a 575 mm long gauge.

Sine bar

The sine bar,[21] shown in Figure 13.16, is used in conjunction with gauge blocks for the calibration of angle-measuring equipment. It consists of a rectangular-section piece of steel into which a pair of rollers are located. The distance between the centres of the rollers is known very accurately and is commonly set to be 250 mm or some multiple thereof. By setting the bar up with one roller on a reference table and the other resting on a pile of gauge blocks, any angle can be set up, given by (see Figure 13.16)

$$\theta = \sin^{-1}(H/L)$$

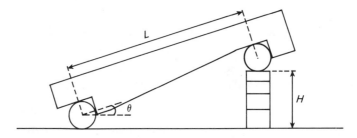

Figure 13.16 Sine bar

13.4 Calibration procedures

The general principles of calibration are to set up a series of standard lengths using gauge blocks and/or length bars and to compare the output reading of the instrument being calibrated against the correct value when the instrument is set to the standard lengths. This comparison is carried out on a surface plane or table and various ancillary blocks (see Figure 13.2) and clamps are used to hold the set of gauge blocks and length bars in a manner appropriate to the needs of the instrument being calibrated. Various other special comments about the calibration of particular instruments can be made as follows.

New *micrometers* are set up with a setting gauge and this instrument may often also be suitable for re-calibrating workshop micrometers. For calibration of working and reference standard micrometers, gauge blocks are used. When using gauge blocks for calibration, it is important that measurement accuracy is checked at intermediate positions of the thimble as well as for full revolutions. A set of gauge blocks set up to give the following dimensions in turn would satisfy this requirement: 2.5 mm, 5.1 mm, 7.7 mm, 12.9 mm, 15.0 mm, 17.6 mm, 20.2 mm, 22.8 mm, 25.0 mm.

As well as checking length-measurement accuracy, calibration checks on micrometers must include checks on the flatness and parallelism of the anvil faces. Flatness is checked by bringing the anvil into contact with an optically flat glass plate. The degree of non-flatness is indicated by the colour and number of interference bands on the surface. Parallelism of the measuring faces is tested by moving an optical flat about between them and observing the changes in the number of interference bands.

If calibration checks show up errors in length measurement, micrometers are provided with a means of adjustment to take up small amounts of wear in the screw thread. They also usually have some means of rotating the body scale so that the zero mark can be re-set.

In the calibration of *height gauges*, *depth gauges* and *dial gauges*, comparison with reference piles of gauge blocks is effected by a special form of dial gauge which has a measurement resolution of 1 μm. In other procedures, suitable combinations of gauge blocks are wrung together and then clamped using special end fittings. These provide the standard reference lengths used in the calibration of *micrometers* and *calipers*.

A reference surface plane is also required for the purpose of calibrating angle-measuring instruments. The calibration instrument, a sine bar, is used in conjunction with a pile of gauge blocks to set up a series of standard angles. The output reading of the instrument under test can then be compared with the correct value as the instrument is set to each of these angles in turn. Again, various blocks and clamps are often used on the surface plane to assist in this procedure.

13.5 Calibration frequency

As with so many other types of measuring instrument, experimentally derived knowledge is needed to define suitable calibration frequencies for dimension-measuring equipment. The required frequency depends largely upon their rate and conditions of usage. However, whatever re-calibration frequency is defined, a vigilant watch must be kept for signs of physical damage to the instrument at intermediate times. Instruments are often abused and mistreated, for instance by using the end of a ruler to open tins and using a vernier caliper as a spanner. Whenever such misuse is suspected, the instrument must be withdrawn immediately for calibration checks to be carried out.

Volume flow-rate calibration

Volume flow-rate measurement is extremely important in all the process industries. It is used for quantifying the flow of all materials which are in a gaseous, liquid or semi-liquid slurry form and carried in pipes.

14.1 Review of instruments measuring volume flow rate

A wide range of instruments have been developed for measuring volume flow rate and they can be divided into the following classes:

   (a) differential pressure meters;
   (b) variable area meters;
   (c) positive displacement meters;
   (d) turbine flowmeters;
   (e) electromagnetic flowmeters;
   (f) vortex shedding flowmeters;
   (g) gate-type meters;
   (h) ultrasonic flowmeters;
   (i) cross-correlation flowmeters;
   (j) laser–Doppler flowmeters.

The number of relevant factors to be considered when specifying a flowmeter for a particular application is very large. These include the temperature and pressure of the fluid, its density, viscosity, chemical properties and abrasiveness, whether it contains particles, whether it is a liquid or gas, etc. The required performance factors of accuracy, measurement range, acceptable pressure drop, output signal characteristics, reliability and service life must also be assessed.

Differential pressure meters

Differential pressure meters involve the insertion of some device into a fluid-carrying pipe which causes an obstruction and creates a pressure difference on either side of the device. The orifice plate is very much the most common of such devices and accounts for 95 per cent of all differential pressure instruments. Other devices in this category include the Venturi tube, the flow nozzle, the Dall flow tube and the Pitot tube. When such a restriction is placed in a pipe, the velocity of the fluid through the restriction increases and the pressure decreases. The volume flow rate is then proportional to the square root of the pressure difference across the obstruction. The manner in which this pressure difference is measured is important. Measuring the two pressures with different instruments and calculating the difference between the two measurements is not satisfactory because of the large measurement error which can arise when the pressure difference is small. The normal procedure is therefore to use a diaphragm-based differential pressure transducer.

All applications of this method of flow measurement assume that flow conditions upstream of the obstruction device are in steady state, and a certain minimum length of straight run of pipe ahead of the flow measurement point is specified to ensure this. The minimum lengths required for various pipe diameters are specified in British Standards tables (and also in alternative but equivalent national standards used in other countries), but a useful rule of thumb widely used in the process industries is to specify a length of ten times the pipe diameter. If physical restrictions make this impossible to achieve, special flow smoothing vanes can be inserted immediately ahead of the measurement point.

Flow-restriction-type instruments are popular because they have no moving parts and are therefore robust, reliable and easy to maintain. One disadvantage of this method is that the obstruction causes a permanent loss of pressure in the flowing fluid. The magnitude and hence importance of this loss depends on the type of obstruction element used, but where the pressure loss is large, it is sometimes necessary to recover the lost pressure by an auxiliary pump further down the flow line. This class of device is not normally suitable for measuring the flow of slurries as the tappings into the pipe to measure the differential pressure are prone to blockage, although the Venturi tube can be used to measure the flow of dilute slurries.

Figure 14.1 illustrates approximately the way in which the flow pattern is interrupted when an orifice plate is inserted into a pipe. The other obstruction devices also have a similar effect. Of particular interest is the fact that the minimum cross-sectional area of flow occurs not within the obstruction but at a point downstream of it. Knowledge of the pattern of pressure variation along

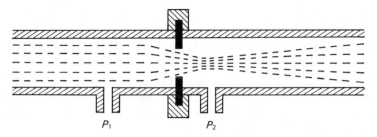

Figure 14.1 Profile of flow across orifice plate

the pipe, as shown in Figure 14.2, is also of importance in using this technique of flow measurement. This shows that the point of minimum pressure coincides with the point of minimum cross-section flow, a little way downstream of the obstruction. Figure 14.2 also shows that there is a small rise in pressure immediately before the obstruction. It is therefore important not only to position the instrument measuring P_2 exactly at the point of minimum pressure, but also to measure the pressure P_1 at a point upstream of the point where the pressure starts to rise before the obstruction.

In the absence of any heat transfer mechanisms, and assuming frictionless flow of an incompressible fluid through the pipe, the theoretical volume flow rate of the fluid, Q, is given by[1]

$$Q = \frac{A_2}{[1 - (A_2/A_1)^2]^{1/2}} \cdot [2(P_1 - P_2)/\rho]^{1/2} \tag{14.1}$$

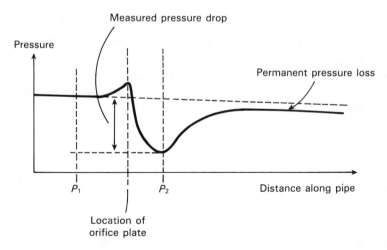

Figure 14.2 Pattern of pressure variation along pipe obstructed by an orifice plate

where A_1 and P_1 are the cross-sectional area and pressure of the fluid flow before the obstruction, A_2 and P_2 are the cross-sectional area and pressure of the fluid flow at the narrowest point of the flow beyond the obstruction, and ρ is the fluid density.

Equation (14.1) is never applicable in practice for several reasons. Firstly, frictionless flow is never achieved. However, in the case of turbulent flow through smooth pipes, friction is low and it can be adequately accounted for by a variable called the Reynolds number, which is a measurable function of the flow velocity and the viscous friction. The other reasons for the non-applicability of equation (14.1) are that the initial cross-sectional area of the fluid flow is less than the diameter of the pipe carrying it and that the minimum cross-sectional area of the fluid is less than the diameter of the obstruction. Therefore, neither A_1 nor A_2 can be measured. These problems are taken account of by modifying equation (14.1) to the following:

$$Q = \frac{C_D \cdot A_2'}{[1 - (A_2'/A_1')^2]^{1/2}} \cdot [2(P_1 - P_2)/\rho]^{1/2} \tag{14.2}$$

where A_1' and A_2' are the pipe diameters before and at the obstruction and C_D is a constant, known as the discharge coefficient, which accounts for the Reynolds number and the difference between the pipe and flow diameters.

Before equation (14.2) can be evaluated, the discharge coefficient must be calculated. As this varies between each measurement situation, it would appear at first sight that the discharge coefficient must be determined by practical experimentation in each case. However, provided that certain conditions are met, standard tables can be used to obtain the value of the discharge coefficient appropriate to the pipe diameter and fluid involved.

It is particularly important in applications of flow-restriction methods to choose an instrument whose range is appropriate to the magnitudes of flow rate being measured. This requirement arises because of the square-root type of relationship between the pressure difference and the flow rate, which means that as the pressure difference decreases, the error in flow-rate measurement can become very large. In consequence, restriction-type flowmeters are only suitable for measuring flow rates between 30 per cent and 100 per cent of the instrument range.

Orifice plate

The orifice plate is a metal disc with a hole in it, as shown in Figure 14.3, inserted into the pipe carrying a flowing fluid. This hole is normally concentric with the disc. Over 50 per cent of the instruments used in industry for

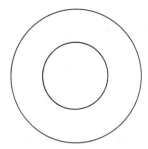

Figure 14.3 The orifice plate

measuring volume flow rate are of this type. The use of the orifice plate is so widespread because of its simplicity, cheapness and availability in a wide range of sizes. However, the best accuracy obtainable with this type of obstruction device is only ±2 per cent and the permanent pressure loss caused in the flow is very high, being between 50 per cent and 90 per cent of the pressure difference $(P_1 - P_2)$ in magnitude. Other problems with the orifice plate are a gradual change in the discharge coefficient over a period of time as the sharp edges of the hole wear away, and a tendency for any particles in the flowing fluid to stick behind the hole and gradually build up and reduce its diameter. The latter problem can be minimized by using an orifice plate with an eccentric hole. If this hole is close to the bottom of the pipe, solids in the flowing fluid tend to be swept through, and build-up of particles behind the plate is minimal.

A very similar problem arises if there are any bubbles of vapour or gas in the flowing fluid when liquid flow is involved. These also tend to build up behind an orifice plate and distort the pattern of flow. This difficulty can be avoided by mounting the orifice plate in a vertical run of pipe.

Flow nozzle

The form of a flow nozzle is shown in Figure 14.4. This is not prone to the problem of solid particles or bubbles of gas in a flowing fluid sticking in the flow restriction, and so in this respect it is superior to the orifice plate. Its useful working life is also greater because it does not become worn away in the same way as an orifice plate. These factors contribute to giving the instrument a greater measurement accuracy and a need for calibration at less frequent intervals. However, as the engineering effort involved in fabricating a flow nozzle is greater than that required to make an orifice plate, the instrument is somewhat more expensive. In terms of the permanent pressure loss imposed on the measured system, the flow nozzle is very similar to the orifice plate. A typical application of the flow nozzle is in the measurement of steam flow.

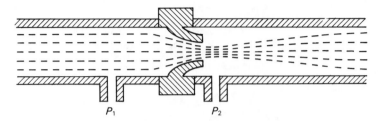

Figure 14.4 The flow nozzle

Venturi

The Venturi is a precision-engineered tube of a special shape, as shown in Figure 14.5. It is a very expensive instrument but offers very good accuracy and imposes a permanent pressure loss on the measured system of only 10–15 per cent of the pressure difference ($P_1 - P_2$) across it. The smooth internal shape of this type of restriction means that it is unaffected by solid particles or gaseous bubbles in the flowing fluid, and in fact can even cope with dilute slurries. It has almost no maintenance requirements and its working life is very long.

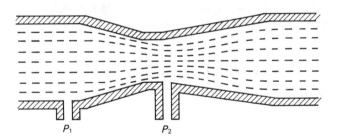

Figure 14.5 The Venturi

Dall flow tube

The Dall flow tube, shown in Figure 14.6, consists of two conical reducers inserted into the fluid-carrying pipe. It has a very similar internal shape to the Venturi, except that it lacks a throat. This construction is much easier to manufacture than a Venturi (which requires complex machining), and this gives the Dall flow tube an advantage in cost over the Venturi, although the measurement accuracy obtained is not quite as good. Another advantage of the Dall

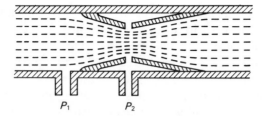

Figure 14.6 The Dall flow tube

flow tube is its shorter length, which makes the engineering task of inserting it into the flow line easier. The Dall tube has one further operational advantage, in that the permanent pressure loss imposed on the system is only about 5 per cent of the measured pressure difference $(P_1 - P_2)$, and thus is only half that due to a Venturi. In other respects, the two instruments are very similar in their low maintenance requirement and long life.

Pitot tube

The Pitot static tube is mainly used for making temporary measurements of flow, although it is also used in some instances for permanent flow monitoring. The instrument depends on the principle that a tube placed with its open end in a stream of fluid, as shown in Figure 14.7, will bring to rest that part of the fluid which impinges on it, and the loss of kinetic energy will be converted to a measurable increase in pressure inside the tube.

The flow velocity can be calculated from the formula

$$v = C[2g(P_1 - P_2)]^{1/2}$$

The constant C, known as the Pitot tube coefficient, is a factor which corrects

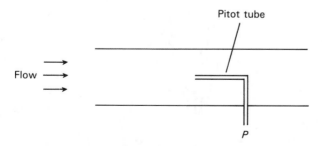

Figure 14.7 The Pitot tube

for the fact that not all fluid incident on the end of the tube will be brought to rest: a proportion will slip around it according to the design of the tube.

Having calculated v, the volume flow rate can then be calculated by multiplying v by the cross-sectional area of the flow pipe, A.

Inferring the volume flow rate from measurement of the flow velocity at one point in the fluid obviously requires the flow profile to be very uniform. If this condition is not met, multiple Pitot tubes can be used to measure velocities across the whole pipe cross-section.

Pitot tubes have the advantage that they cause negligible pressure loss in the flow. They are also cheap, and the installation procedure consists of the very simple process of pushing them down a small hole drilled in the flow-carrying pipe.

Their main failing is that the lowest measurement uncertainty achievable is usually about ±5 per cent, and sensitive pressure-measuring devices are needed to achieve even this limited level of accuracy, as the pressure difference created is very small. More recently, measurement capabilities with the uncertainty down to ±1 per cent have been claimed for specially designed Pitot tubes.[2]

Variable-area flowmeters

In this class of flowmeter, the differential pressure across a variable aperture is used to adjust the area of the aperture. The aperture area is then a measure of

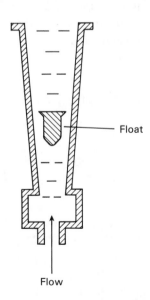

Float

Flow

Figure 14.8 The variable-area flowmeter

the flow rate. This type of instrument only gives a visual indication of flow rate, and so is of no use in automatic control schemes. However, it is reliable, cheap and used extensively throughout industry, accounting for 20 per cent of all flowmeters sold.

In its simplest form, shown in Figure 14.8, the instrument consists of a tapered glass tube containing a float which takes up a stable position where its submerged weight is balanced by the upthrust due to the differential pressure across it. The position of the float is a measure of the effective annular area of the flow passage and hence of the flow rate. The accuracy of the cheapest instruments is only ±3 per cent, but more expensive versions offer measurement accuracies as high as ±0.2 per cent. The normal measurement range is between 10 per cent and 100 per cent of the full-scale reading for any particular instrument.

Positive-displacement flowmeters

All positive-displacement meters operate by using mechanical divisions to displace discrete volumes of fluid successfully. While this principle of operation is common, many different mechanical arrangements exist for putting the principle into practice. All versions of positive-displacement meter are low-friction, low-maintenance and long-life devices, although they do impose a small permanent pressure loss on the flowing fluid. Low friction is especially important when measuring gas flows, and meters with special mechanical arrangements to satisfy this requirement have been developed.

The rotary piston meter illustrated in Figure 14.9, is the most common type of positive-displacement meter. This uses a cylindrical piston which is displaced

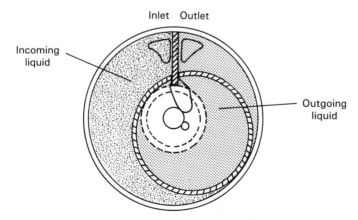

Figure 14.9 Rotary piston form of positive-displacement flowmeter

around a cylindrical chamber by the flowing fluid. Rotation of the piston drives an output shaft. This can either be used with a pointer-and-scale system to give a visual flow reading or it can be converted into an electrical output signal.

Positive-displacement flowmeters account for nearly 10 per cent of the total number of flowmeters used in industry. Such devices are used in large numbers for metering domestic gas and water consumption and they are also capable of measuring the flow of high-viscosity fluids. The cheapest instruments have an accuracy of about ±1.5 per cent, but the accuracy in more expensive instruments can be as good as ±0.2 per cent. These higher-quality instruments are used extensively within the oil industry, as such applications can justify the high cost of such instruments.

Turbine meters (inferential meters)

A turbine or inferential flowmeter consists of a multi-bladed wheel mounted in a pipe along an axis parallel to the direction of fluid flow in the pipe, as shown in Figure 14.10. The flow of fluid past the wheel causes it to rotate at a rate which is proportional to the volume flow rate of the fluid. This rate of rotation is measured by constructing the flowmeter such that it behaves as a variable

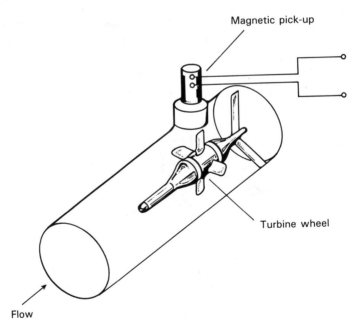

Figure 14.10 The turbine flowmeter

reluctance tachogenerator. This is achieved by fabricating the turbine blades from a ferromagnetic material and placing a permanent magnet and coil inside the meter housing. A voltage pulse is induced in the coil as each blade on the turbine wheel moves past it, and if these pulses are measured by a pulse counter, the pulse frequency, and hence flow rate, can be deduced. Provided that the turbine wheel is mounted in low-friction bearings, measurement uncertainty can be as low as ±0.1 per cent. However, turbine flowmeters are less rugged and reliable than flow-restriction-type instruments, and are badly affected by any particulate matter in the flowing fluid. Bearing wear is a particular problem and turbine meters also impose a permanent pressure loss on the measured system.

Turbine meters are particularly prone to large errors when there is any significant second phase in the fluid measured. For instance, using a turbine meter calibrated on pure liquid to measure a liquid containing 5 per cent air produces a 50 per cent measurement error. As an important application of the turbine meter is in the petrochemical industries, where gas–oil mixtures are common, special procedures are being developed to avoid such large measurement errors. The most promising approach is to homogenize the two gas–oil phases prior to flow measurement.[3]

Turbine meters have a similar cost and market share to positive displacement meters, and compete for many applications, particularly in the oil industry. Turbine meters are smaller and lighter than the latter and are preferred for low-viscosity, high-flow measurements. Positive-displacement meters are superior, however, in conditions of high viscosity and low flow rate.

Electromagnetic flowmeters

Electromagnetic flowmeters are limited to measuring the volume flow rate of electrically conductive fluids and account for about 5 per cent of the flowmeters used in industry. A reasonable measurement accuracy is given, with uncertainty levels typically around ±1.0 per cent, but the instrument is expensive to buy. A further significant expense is the need for careful calibration of each instrument individually during manufacture, as there is considerable variation in the properties of the magnetic materials used. Running costs in terms of electricity consumption are also generally high, although versions with a much lower power consumption (and ±0.5 per cent inaccuracy) are now becoming available.[4]

The instrument, shown in Figure 14.11, consists of a stainless steel cylindrical tube, fitted with an insulating liner, which carries the measured fluid. Typical lining materials used are neoprene, polytetrafluoroethylene (PTFE) and polyurethane. A magnetic field is created in the tube by placing mains-energized field coils on either side of it, and the voltage induced in the fluid is measured

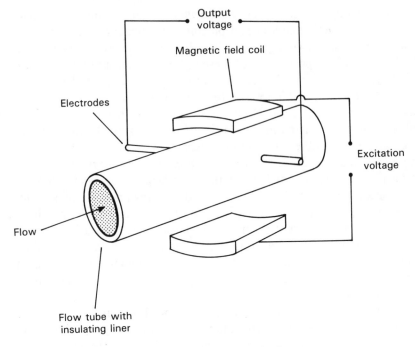

Figure 14.11 The electromagnetic flowmeter

by two electrodes inserted into opposite sides of the tube. The ends of these electrodes are usually flush with the inner surface of the cylinder. The electrodes are constructed from a material which is unaffected by most types of flowing fluid, such as stainless steel, platinum–iridium alloys, hastelloy, titanium and tantalum. In the case of the rarer metals in this list, the electrodes account for a significant part of the total instrument cost.

By Faraday's law of electromagnetic induction, the voltage, E, induced across a length, L, of the flowing fluid moving at velocity, v, in a magnetic field of flux density, B, is given by

$$E = B \cdot L \cdot v \tag{14.3}$$

L is the distance between the electrodes, which is the diameter of the tube, and B is a known constant. Hence, measurement of the voltage E induced across the electrodes allows the flow velocity v to be calculated from equation (14.3). Having thus calculated v, it is a simple matter to multiply v by the cross-sectional area of the tube to obtain a value for the volume flow rate. The typical voltage signal measured across the electrodes is 1 mV when the fluid flow rate is 1 m/s.

The internal diameter of a magnetic flowmeter is normally the same as that of the rest of the flow-carrying pipework in the system. Therefore, there is no obstruction to the fluid flow and consequently no pressure loss associated with measurement. Like other forms of flowmeter, the magnetic type requires a minimum length of straight pipework immediately prior to the point of flow measurement in order to guarantee the accuracy of measurement, although a length equal to five pipe diameters is usually sufficient.

While the flowing fluid must be electrically conductive, the method is of use in many applications and is particularly useful for measuring the flow of slurries in which the liquid phase is electrically conductive. It is also suitable for measuring the flow of many types of corrosive fluid. At the present time, magnetic flowmeters account for about 15 per cent of the new flowmeters sold, and this total is slowly growing. One operational problem is that the insulating lining is subject to damage when abrasive fluids are being handled, and this can give the instrument a limited life.

Current new developments in electromagnetic flowmeters are producing physically smaller instruments and employing better coil designs which reduce electricity consumption. Also, whereas conventional electromagnetic flowmeters require a minimum fluid conductivity of 10 μmho/cm^3, some new versions are now becoming available which can cope with fluid conductivities as low as 1 μmho/cm^3.

Vortex-shedding flowmeters

Vortex-shedding flowmeters only account for about 1 per cent of flowmeters sold at present, but this percentage is likely to grow in the future as the characteristics of these instruments become more generally known. The operating principle of the instrument is based on the natural phenomenon of vortex shedding, created by placing an unstreamlined obstacle (known as a bluff body) in a fluid-carrying pipe, as indicated in Figure 14.12. When fluid flows past the obstacle, boundary layers of viscous, slow-moving fluid are formed along the outer surface. Because the obstacle is not streamlined, the flow cannot follow the contours of the body on the downstream side, and the separate layers become detached and roll into eddies or vortices in the low-pressure region behind the obstacle. The shedding frequency of these alternately shed vortices is proportional to the fluid velocity past the body. Various thermal, magnetic, ultrasonic and capacitive vortex detection techniques are employed in different instruments.

Such instruments have no moving parts, operate over a wide flow range, have a low power consumption and require little maintenance. They can measure both liquid and gas flows and a common inaccuracy figure quoted is

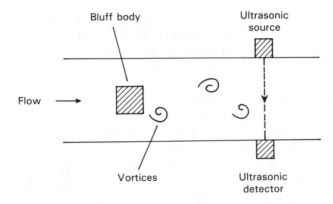

Figure 14.12 The vortex shedding flowmeter

±1 per cent of full-scale reading, though this can be seriously downgraded in the presence of flow disturbances upstream of the measurement point, and a straight run of pipe before the measurement point of fifty pipe diameters is recommended. Another problem with the instrument is its susceptibility to pipe vibrations, although new designs which have a better immunity to such vibrations are becoming available.

Gate-type meters

The *gate meter* was the earliest device in this class. It consists of a spring-loaded, hinged flap mounted at right-angles to the direction of fluid flow in the fluid-carrying pipe. The flap is connected to a pointer outside the pipe. The fluid flow deflects the flap and pointer and the flow rate is indicated by a graduated scale behind the pointer. The major difficulty with such devices is in preventing leaks at the hinge point.

A variation on this principle is the *air-vane meter* which measures deflection of the flap by a potentiometer inside the pipe. This is commonly used to measure air flow within automotive fuel-injection systems.

Another device in this class is the *target meter*. This consists of a circular-disc-shaped flap in the pipe. Fluid flow rate is inferred from the force exerted on the disc measured by strain gauges bonded to it. The meter is very useful for measuring the flow of dilute slurries but it does not find wide application elsewhere as it has a relatively high cost.

Ultrasonic flowmeters

The ultrasonic technique of volume flow-rate measurement is, like the magnetic flowmeter, a non-invasive method. It is not restricted to conductive fluids, however, and is particularly useful for measuring the flow of corrosive fluids and slurries. A further advantage over magnetic flowmeters is that the instrument is one which clamps on externally to existing pipework rather than being inserted as an integral part of the flow line, as in the case of the magnetic flowmeter. As the procedure of breaking into a pipeline to insert a flowmeter can be as expensive as the cost of the flowmeter itself, the ultrasonic flowmeter has significant cost advantages. Its clamp-on mode of operation has safety advantages in avoiding the possibility of personnel installing flowmeters coming into contact with hazardous fluids such as poisonous, radioactive, flammable or explosive fluids. Also, any contamination of the fluid being measured (e.g. food substances and drugs) is avoided. The introduction of this type of flowmeter is a comparatively recent one and its present market share is very small. In view of its distinct advantages, however, its industrial use is likely to increase over the next few years.

Two different types of ultrasonic flowmeter exist which employ distinct technologies, based on Doppler shift and on transit time. In the past, this has not always been readily understood, and has resulted in ultrasonic technology being rejected entirely when one of these two forms has been found to be unsatisfactory in a particular application. This is unfortunate because the two technologies have distinct characteristics and areas of application, and many situations exist where one form is very suitable and the other not suitable. To reject both, having only tried out one, is therefore a serious mistake. Rough guidelines about which type to choose could be stated as follows: for clean liquids use transit-time types and for gritty, aerated liquids use Dopler-shift types.

Particular care has to be taken to ensure a stable flow profile in ultrasonic flowmeter applications. It is usual to increase the normal specification of the minimum length of straight pipe run prior to the point of measurement, expressed as a number of pipe diameters, from a figure of ten up to twenty, or in some cases even fifty, diameters. Analysis of the reasons for poor performance in many instances of ultrasonic flowmeter application has shown failure to meet this stable flow-profile requirement to be a significant factor.

Doppler-shift ultrasonic flowmeter

The principle of operation of the Doppler-shift flowmeter is shown in Figure 14.13. A fundamental requirement of these instruments is the presence of

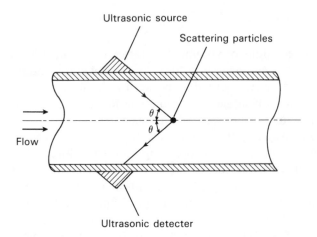

Figure 14.13 Doppler-shift ultrasonic flowmeter

scattering elements within the flowing fluid which deflect the ultrasonic energy output from the transmitter such that it enters the receiver. These can be provided by either solid particles, gas bubbles or eddies in the flowing fluid. The scattering elements cause a frequency shift between the transmitted and reflected ultrasonic energy, and measurement of this shift enables the fluid velocity to be inferred.

The instrument consists essentially of an ultrasonic transmitter–receiver pair clamped onto the outside wall of a fluid-carrying vessel. Ultrasonic energy consists of a train of short bursts of sinusoidal waveforms at a frequency between 0.5 MHz and 20 MHz. This frequency range is described as ultrasonic because it is outside the range of human hearing. The flow velocity, v, is given by

$$v = \frac{c(f_t - f_r)}{2 \cdot f_t \cdot \cos(\theta)} \qquad (14.4)$$

where f_t and f_r are the frequencies of the transmitted and received ultrasonic waves, respectively, c is the velocity of sound in the fluid being measured, and θ is the angle that the incident and reflected energy waves make with the axis of flow in the pipe. Volume flow rate is then readily calculated by multiplying the measured flow velocity by the cross-sectional area of the fluid-carrying pipe.

The electronics involved in Doppler-shift flowmeters is relatively simple and therefore cheap. Ultrasonic transmitters and receivers are also relatively inexpensive, being based on piezo-electric oscillator technology. As all its components are cheap, the Doppler-shift flowmeter itself is inexpensive. The measurement accuracy obtained depends on many factors such as the flow profile, the constancy of pipe-wall thickness, the number, size and spatial distribution of scatterers, and the accuracy with which the speed of sound in the fluid is known. Consequently, accurate measurement can only be achieved by the tedious procedure of carefully calibrating the instrument in each particular flow-measurement application. Otherwise, measurement errors can approach ±10 per cent of the reading, and for this reason Doppler-shift flowmeters are often used merely as flow indicators, rather than for accurate quantification of the volume flow rate.

Versions are now available which avoid the problem of variable pipe thickness by being fitted inside the flow pipe, flush with its inner surface. Accuracy of ±0.5 per cent is claimed for such devices. Other recent developments are the use of multiple-path ultrasonic flowmeters which use an array of ultrasonic elements to obtain an average velocity measurement which substantially reduces the error due to non-uniform flow profiles. There is a substantial cost penalty involved in this, however.

Transit-time ultrasonic flowmeter

The transit-time ultrasonic flowmeter is an instrument designed for measuring the volume flow rate in clean liquids or gases. It consists of a pair of ultrasonic transducers mounted along an axis aligned at an angle θ with respect to the fluid-flow axis, as shown in Figure 14.14. Each transducer consists of a transmitter–receiver pair, with the transmitter emitting ultrasonic energy which

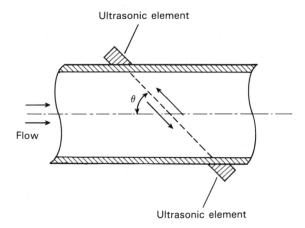

Figure 14.14 Transit-time ultrasonic flowmeter

travels across to the receiver on the opposite side of the pipe. These ultrasonic elements are normally piezo-electric oscillators of the same type as are used in Doppler-shift flowmeters. Fluid flowing in the pipe causes a time difference between the transit times of the beams travelling upstream and downstream, and measurement of this difference allows the flow velocity to be calculated. The typical magnitude of this time difference is 100 ns in a total transit time of 100 μs, and high-precision electronics are therefore needed to measure it. There are three distinct ways of measuring the time shift: direct measurement; conversion to a phase change; and conversion to a frequency change. The third of these options is particularly attractive as it obviates the need to measure the speed of sound in the measured fluid which the first two methods require. A scheme applying this third option is shown in Figure 14.15. This also multiplexes the transmitting and receiving functions, so that only one ultrasonic element is needed in each transducer.

The forward and backward transit times across the pipe, T_f and T_b, are given by

$$T_f = \frac{L}{c + v \cdot \cos(\theta)} \quad ; \quad T_b = \frac{L}{c - v \cdot \cos(\theta)}$$

where c is the velocity of sound in the fluid, v is the flow velocity, L is the distance between the ultrasonic transmitter and receiver, and θ is the angle of the ultrasonic beam with respect to the fluid-flow axis. The time difference ΔT is given by

$$\Delta T = T_b - T_f = \frac{2 \cdot v \cdot L \cdot \cos(\theta)}{c^2 - v^2 \cdot \cos^2(\theta)}$$

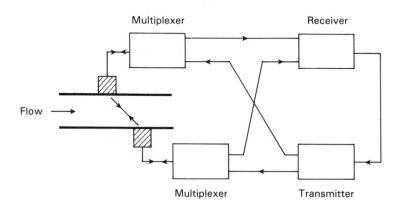

Figure 14.15 Transit-time measurement system

This requires knowledge of c before it can be solved. However, a solution can be found much more simply if the receipt of a pulse is used to trigger the transmission of the next ultrasonic energy pulse. Then, the frequencies of the forward and backward pulse trains are given by

$$F_f = \frac{1}{T_f} = \frac{c - v \cdot \cos(\theta)}{L} \quad ; \quad F_b = \frac{1}{T_b} = \frac{c + v \cdot \cos(\theta)}{L}$$

If the two frequency signals are now multiplied together, the resulting beat frequency is given by

$$\Delta F = F_b - F_f = \frac{2 \cdot v \cdot \cos(\theta)}{L}$$

c has now been eliminated and v can be calculated from a measurement of ΔF as

$$v = \frac{L \cdot \Delta F}{2 \cdot \cos(\theta)}$$

Transit-time flowmeters are of more general use than Doppler-shift flowmeters, particularly where the pipe diameter involved is large and hence the transit time is consequently sufficiently large to be measured with reasonable accuracy. It is possible to achieve an accuracy figure of ± 0.5 per cent. The instrument costs more than a Doppler-shift flowmeter, however, because of the greater complexity of the electronics needed to make accurate transit-time measurements.

Cross-correlation flowmeters

The cross-correlation flowmeter is a type of flowmeter which has not yet achieved widespread practical use in industry. Much development work is still going on, and the instruments thus mainly exist only as prototypes in research laboratories. However, they are included here because their use is likely to become much more widespread in the future. Cost is likely to be similar to that of the larger versions of electromagnetic flowmeter.

Such instruments require some detectable random variable to be present in the flowing fluid. This can take forms such as velocity turbulence and temperature fluctuations. When such a stream of variables is detected by a sensor, the output signal generated consists of noise with a wide frequency spectrum.

Cross-correlation flowmeters use two such sensors placed a known distance apart in the fluid-carrying pipe and cross-correlation techniques are applied to the two output signals from these sensors. This procedure compares one signal with progressively time-shifted versions of the other signal until the best match is obtained between the two waveforms. If the distance between the sensors is divided by this time shift, a measurement of the flow velocity is obtained. A digital processor is an essential requirement to calculate the cross-correlation function, and therefore the instrument must be properly described as an intelligent one. Measurement uncertainty is typically ±2 per cent.

In practice, the existence of random disturbances in the flow is unreliable, and their detection is difficult. To answer this problem, ultrasonic cross-correlation flowmeters are under development. These use ultrasonic transducers to inject disturbances into the flow and also to detect the disturbances further downstream.

A particular application which is likely to develop for cross-correlation flowmeters is in measuring multi-phase flow where the fluid is a mixture of two or more gas, liquid and solid components. Successful measurement of solid–liquid flows has been demonstrated and trials with gas–liquid flows are encouraging.[5]

Further information about cross-correlation flowmeters can be found in reference 6.

Laser Doppler flowmeter

This instrument gives direct measurements of flow velocity for liquids containing suspended particles flowing in a transparent pipe. Light from a laser is focussed by an optical system to a point in the flow. The movement of particles causes a Doppler shift of the scattered light and produces a signal in a photo-detector which is related to the fluid velocity. A very wide range of flow velocities between 10 μm/s and 800 m/s can be measured by this technique.

Sufficient particles for satisfactory operation are normally present naturally in most liquid and gaseous fluids, and the introduction of artificial particles is rarely needed. The technique is advantageous in measuring flow velocity directly rather than inferring it from a pressure difference. It also causes no interruption in the flow and, as the instrument can be made very small, it can measure velocity in confined areas. One limitation is that it measures local flow velocity in the vicinity of the focal point of the light beam, which can lead to large errors in the estimation of mean volume flow rate if the flow profile is not uniform. However, this limitation is often used constructively in applications of the instrument where the flow profile across the cross-section of a pipe is determined by measuring the velocity at a succession of points.

Intelligent flowmeters

All the usual benefits associated with intelligent instruments are potentially applicable to many types of flowmeter. One general benefit in flowmeters is extension of the measurement range for a particular instrument while maintaining a good standard of accuracy. However, the availability of intelligent flowmeters in the market place is currently very limited.

Intelligent differential pressure-measuring instruments can be used to good effect in conjunction with obstruction-type flow transducers. One immediate benefit of this in the case of the commonest flow-restriction device, the orifice plate, is to extend the lowest flow measurable with acceptable accuracy down to 20 per cent of the maximum flow value.

In positive-displacement meters, intelligence allows compensation for thermal expansion of meter components and temperature-induced viscosity changes. Correction for variations in flow pressure is also provided for.

Intelligent electromagnetic flowmeters are also available, and these have a self-diagnosis and self-adjustment capability. The usable instrument range is typically from 3 per cent to 100 per cent of the full-scale reading and quoted accuracy is ±0.5 per cent. It is also normal to include a non-volatile memory to protect constants used for correcting for modifying inputs, etc., against power supply failures.

Intelligent turbine meters are able to detect their own bearing wear and also report deviations from initial calibration due to blade damage, etc. Some versions also have self-adjustment capability.

The trend is now moving towards total flow computers which can process inputs from almost any type of transducer. Such devices allow user-input of parameters such as specific gravity, fluid density, viscosity, pipe diameters, thermal expansion coefficients, discharge coefficients, etc. Auxiliary inputs from temperature transducers are also catered for. After processing the raw flow transducer output with this additional data, flow computers are able to produce measurements of flow to a very high degree of accuracy.

14.2 Introduction to calibration

Calibration of volume-flow-measuring instruments is a relatively expensive procedure even when only a moderate level of accuracy is demanded. Where high accuracy is required, the cost of calibration can be very great indeed. Therefore, in quality control systems which involve flow measurements, it is particularly important to establish exactly what accuracy level is needed so that the calibration system instituted does not cost more than necessary. In some cases, such as handling valuable fluids or where there are legal requirements as in petrol pumps, high accuracy levels (e.g. error $\leqslant 0.1$ per cent) are justified. In other situations, however, such as in measuring additives to the main stream in a process plant, only low levels of accuracy are needed (e.g. error ≈ 5 per cent).

It is the normal practice to calibrate process flow-measuring instruments on-site as far as possible. This ensures that calibration is performed in the actual flow conditions, which are difficult or impossible to reproduce exactly in a laboratory. This is necessary because the accuracy of flow measurement is greatly affected by the flow conditions and characteristics of the flowing fluid. Because of this, it is also standard practice to repeat flow calibration checks until the same reading is obtained in two consecutive tests. However, Berman has suggested that even these precautions are inadequate and that statistical procedures are needed.[7]

The equipment and procedures used for calibration depend on whether gaseous or liquid flows are being measured. Therefore, separate sections are devoted to each of these cases. It must also be stressed that all calibration procedures refer only to flows of single-phase fluids (i.e. liquids or gases). Where a second or third phase is present (i.e. there is a mixture of gas and/or liquid and/or solid), suitably accurate calibration techniques have yet to be established.[3]

14.3 Calibration equipment and procedures for liquid flow

Calibrated tank

Probably the simplest piece of equipment available for calibrating instruments measuring liquid flow rates is the calibrated tank. This consists of a cylindrical vessel, as shown in Figure 14.16, with conical ends which facilitate draining and cleaning of the tank. A sight tube with a graduated scale is placed alongside the final, upper, cylindrical part of the tank and this allows the volume of liquid in the tank to be measured accurately. Flow-rate calibration is performed by measuring the time taken, starting from an empty tank, for a given volume of liquid to flow into the vessel.

Because the calibration procedure starts and ends in zero-flow conditions, it is not suitable for calibrating instruments which are affected by flow acceleration and deceleration characteristics. This therefore excludes such instruments as differential pressure meters (orifice plate, flow nozzle, Venturi, Dall flow tube, Pitot tube), turbine flowmeters and vortex-shedding flowmeters. The technique is further limited to the calibration of low-viscosity liquid flows, although lining the tank with an epoxy coating can allow the system to cope with somewhat higher viscosities. The limiting factors in this case are the drainage

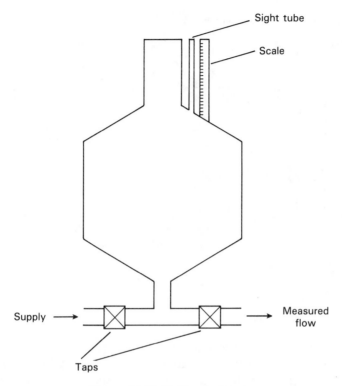

Figure 14.16 Calibrated tank

characteristics of the tank, which must be such that the residue liquid left after draining has an insufficient volume to affect the accuracy of the next calibration.

Pipe prover

The commonest form of pipe prover is the bi-directional type, shown in Figure 14.17, which consists of a U-shaped tube of metal of accurately known cross-section. The purpose of the U-bend is to give a long flow path within a compact spatial volume. Alternative versions with more than one U-bend also exist to cater for situations where an even longer flow path is required. Inside the tube is a hollow, inflatable sphere which is filled with water until its diameter is about 2 per cent larger than that of the tube. As such, the sphere forms a seal with the sides of the tube and acts as a piston. The prover is connected into the existing fluid-carrying pipe network via tappings on either side of a bypass

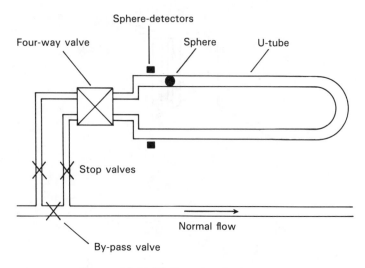

Figure 14.17 Bi-directional pipe prover

valve. A four-way valve at the start of the U-tube allows fluid to be directed in either direction around it. Calibration is performed by diverting flow into the prover and measuring the time taken for the sphere to travel between two detectors in the tube. The detectors are normally of an electromechanical, plunger type.

Uni-directional versions of the above also exist in which fluid only flows in one direction around the tube. A special handling valve has to be provided to return the sphere to the starting point after each calibration, but the absence of a four-way flow control valve makes such devices significantly cheaper than bi-directional types.

Pipe provers are particularly suited to the calibration of pressure-measuring instruments which have a pulse type of output, such as turbine meters. In such cases, the detector switches in the tube can be made to gate the instrument's output pulse counter. This enables not only the basic instrument to be calibrated, but also the ancillary electronics within it at the same time.

The inaccuracy level of such provers can be as low as ± 0.1 per cent. This level of accuracy is maintained for high fluid-viscosity levels and also at very high flow rates. Even higher accuracy is provided by an alternative form of prover which consists of a long, straight metal tube containing a metal piston. However, such devices are more expensive than the other types discussed above and their large space requirements also often cause great difficulties.

Compact prover

The compact prover has an identical operating principle to that of the other pipe provers described above but occupies a much smaller spatial volume. It is therefore used extensively in situations where there is insufficient room to use a larger prover. Many different designs of compact prover exist, operating both in the uni-directional and bi-directional modes, and one such design is shown in Figure 14.18. Common features of compact provers are an accurately machined cylinder containing a metal piston which is driven between two reference marks by the flowing fluid. The instants at which the reference marks are passed are detected by switches, of optical form in the case of the version shown in Figure 14.18. Provision has to be made within these instruments for returning the piston back to the starting point after each calibration and a hydraulic system is commonly used for this. Again, measuring the piston traverse time is made easier if the switches can be made to gate a pulse train, and therefore compact provers are also most suited to instruments having a pulse-type output such as turbine meters. Measurement uncertainty levels down to ±0.1 per cent are possible.

The main technical difficulty in compact provers is measuring the traverse time, which can be as brief as 1 s. The pulse count from a turbine meter in this time would typically be only about 100, making the possible measurement error 1 per cent. To overcome this problem, electronic pulse-interpolation techniques have been developed which can count fractions of pulses.

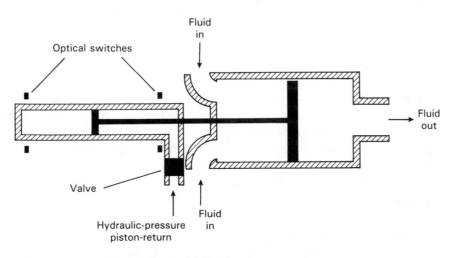

Figure 14.18 Compact prover

Positive-displacement meter

High-quality versions of the positive-displacement flowmeter can be used as a reference standard in flowmeter calibration. The general principles of these were explained in Section 14.1. Such devices can give measurement inaccuracy levels down to ±0.2 per cent.

Gravimetric method

A variation on the principle of measuring the volume of liquid flowing in a given time is to weigh the quantity of fluid flowing in a given time. Apart from its applicability to a wider range of instruments, this technique is not limited to low-viscosity fluids as any residual fluid in the tank before calibration will be detected by the load-cells and therefore compensated for. In the simplest implementation of this system, fluid is allowed to flow for a measured length of time into a tank resting on load-cells. As before, the stop–start mode of fluid flow makes this method unsuitable for calibrating differential pressure, turbine and vortex-shedding flowmeters. It is also unsuitable for measuring high flow rates because of the difficulty in bringing the fluid to rest. These restrictions can be overcome by directing the flowing fluid into the tank via diverter valves. In this alternative, it is important that the timing system be carefully synchronized with the operation of the diverter valves.

All versions of gravimetric calibration equipment are less robust than volumetric types and thus on-site use is not recommended.

Orifice plate

A flow line equipped with a certified orifice plate is sometimes used as a reference standard in flow calibration, especially for high flow rates through large bore pipes. While measurement uncertainty is of the order of ±1 per cent at best, this is adequate for calibrating many flow-measuring instruments.

Turbine meter

Turbine meters are also used as a reference standard for testing flowmeters. Their main application, as for orifice plates, is in calibrating high flow rates through large-bore pipes. Measurement uncertainty down to ±0.2 per cent is attainable.

14.4 Calibration equipment and procedures for gaseous flow

Calibration of gaseous flows poses considerable difficulties compared with calibrating liquid flows. These problems include the lower density of gases, their compressibility and the difficulty of establishing a suitable liquid–air interface as utilized in many liquid flow measurement systems. In consequence, the main methods of calibrating gaseous flows, as described below, are small in number. Certain other specialized techniques, including the gravimetric method and the pressure–volume–temperature method, are also available, as described elsewhere.[8] These provide primary reference standards for gaseous flow calibration with measurement uncertainty down to ±0.3 per cent. However, the expense of the equipment involved is such that it is usually only available in national standards laboratories.

Bell prover

The bell prover consists of a hollow, inverted, metal cylinder suspended over a bath containing light oil, as shown in Figure 14.19. The air volume in the cylinder above the oil is connected, via a tube and a valve, to the flowmeter being calibrated. An air flow through the meter is created by allowing the cylinder to fall downwards into the bath, thus displacing the air contained within it. The flow rate, which is measured by timing the rate of fall of the cylinder, can be adjusted by changing the value of counterweights attached via a low-friction pulley system to the cylinder. This is essentially laboratory-only equipment and therefore on-site calibration is not possible.

Positive-displacement meter

As for liquid flow calibration, positive-displacement flowmeters can be used for the calibration of gaseous flows with inaccuracy levels down to ±0.2 per cent.

Compact prover

Compact provers of the type used for calibrating liquid flows are unsuitable for application to gaseous flows. However, special designs of compact prover are being developed for gaseous flows and hence such devices may find application in gaseous flow calibration in the future.

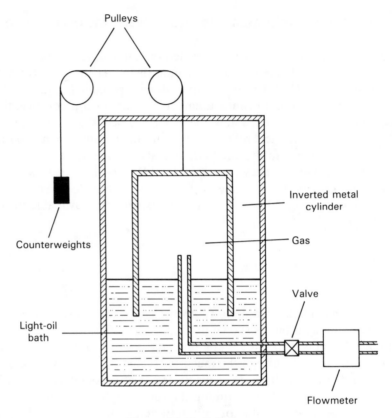

Figure 14.19 Bell prover

14.5 Reference standards

Traceability of flow-rate calibration to fundamental standards is provided for by reference to primary standards of the separate quantities that flow rate is calculated from. Mass measurements are calibrated by comparison with a copy of the international standard kilogram (see Chapter 12) and time is calibrated by reference to a caesium resonator standard. Volume measurements are calibrated against standard reference volumes which are themselves calibrated gravimetrically using a mass-measurement system traceable to the standard kilogram.

Calibration of miscellaneous parameters

15.1 Vibration and Shock

Vibration

One common parameter in the quality evaluation of a product is its ability to withstand vibrations up to a certain magnitude, and therefore a test rig has to be set up in which vibrations are excited and measured. Vibration measurement is also necessary to monitor the operation of products such as machines which can excite vibrations, to check that vibration levels do not exceed limits defined in the quality documentation. Consequently, the calibration of vibration-measuring equipment is an esential part of many quality systems.

Vibration normally consists of linear harmonic motion in which the motion parameters are related by the following formulae:

$$x = X_p \sin(\omega t)$$

$$v = \frac{dx}{dt} = V_p \cos(\omega t)$$

$$a = \frac{dv}{dt} = A_p \sin(\omega t + \pi)$$

(15.1)

where X_p, V_p and A_p are the peak values of the displacement, velocity and acceleration, x, v and a are instantaneous values of these quantities and (ωt) is the angular frequency.

Thus, the intensity of vibration can be measured in terms of either displacement, velocity or acceleration. It is apparent from the above equations that displacements are large at low frequencies and that therefore either displacement or velocity transducers are theoretically best for measuring vibration at such frequencies. Accelerometers would only appear to be preferable at high frequencies. However, there are considerable practical difficulties in mounting and calibrating displacement and velocity transducers and they are rarely used. Consequently, vibration is usually measured by accelerometers at all frequencies. The most common type of transducer used is the piezo-accelerometer, which has typical inaccuracy levels of ±2 per cent.

When measuring vibration, consideration must be given to the fact that attaching an accelerometer to the vibrating body will significantly affect the vibration characteristics if the body has a low mass. The effect of such 'loading' of the measured system, can be quantified by the following equation:

$$a_l = a_b \left[\frac{m_b}{m_b + m_a} \right]$$

where a_l is the acceleration of the body with accelerometer attached, a_b is the acceleration of the body without the accelerometer, m_a is the mass of the accelerometer and m_b is the mass of the body.

Such considerations emphasize the advantage of piezo-accelerometers, as these have a lower mass than other forms of accelerometer and so contribute least to this system-loading effect.

As well as an accelerometer, a vibration measurement system requires other elements, as shown in Figure 15.1, to translate the accelerometer output into a recorded signal. The three other necessary elements are a signal conditioning element, a signal analyser and a signal recorder. The signal conditioning element amplifies the relatively weak output signal from the accelerometer and also transforms the high output impedance of the accelerometer to a lower impedance value. The signal analyser then converts the signal into the form required for output. The output parameter may be either displacement, velocity or acceleration and this may be expressed as either the peak-value, rms-value or average absolute value. The final element of the measurement system is the signal recorder.

All elements of the measurement system, and especially the signal recorder,

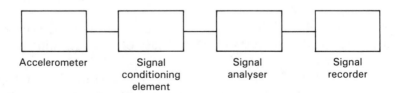

Accelerometer   Signal conditioning element   Signal analyser   Signal recorder

Figure 15.1 Vibration measurement system

must be chosen very carefully to avoid distortion of the vibration waveform. The bandwidth should be such that it is at least a factor of ten better than the bandwidth of the vibration frequency components at both ends. Thus its lowest frequency limit should be less than or equal to 0.1 times the fundamental frequency of vibration and its upper frequency limit should be greater than or equal to ten times the highest significant vibration frequency component.

All elements in the measurement system require calibration, but the element which requires most frequent attention is the vibration transducer, which is normally a piezo-accelerometer. At the lowest level, calibration of the vibration transducer is performed by comparison with a standard accelerometer. In this procedure, both transducers are mounted on a piece of equipment which is able to excite vibrations at a range of frequencies and amplitudes. The standard instrument is most commonly a piezo-electric accelerometer, as this has the best stability and frequency range. If the vibration transducer itself measures acceleration, as is commonly the case, the outputs of the two transducers can be compared directly. Otherwise, the outputs have to be converted to the same forms using equations (15.1) above. Calibration at yearly intervals is usually recommended.

Reference standard accelerometers are similarly used in comparison procedures at all intermediate levels of calibration. Only at the very highest levels of calibration are absolute methods used. The most common absolute methods of calibration involve interferometric techniques which give measurement uncertainty levels of around ±0.5 per cent. Such facilities are only usually found in national standards laboratories.

Shock

Shock describes a type of motion where a moving body is brought suddenly to rest, often because of a collision. This is very common in industrial situations and usually involves a body being dropped and hitting the floor. The ability of a product to withstand shocks of a certain magnitude is often a quality-related parameter and therefore instruments which measure shock, and their calibration, are very important.

Shocks characteristically involve large-magnitude decelerations (e.g. 500 g) which last for a very short time (e.g. 5 ms). An instrument having a very high frequency response is required for shock measurement, and for this reason, piezo-crystal-based accelerometers are commonly used. Again, other elements for analysing and recording the signal are required, as shown in Figure 15.1, and described in the last section. A storage oscilloscope is a suitable instrument for recording the output signal, as this allows the time duration as well as the acceleration levels in the shock to be measured. Alternatively, if a permanent

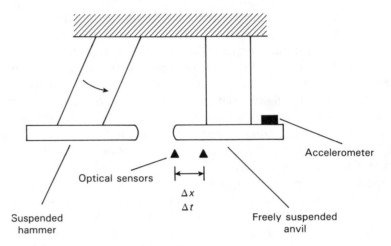

Figure 15.2 Shock machine

record is required, the screen of a standard oscilloscope can be photographed. A further option is to record the output on magnetic tape, which facilitates computerized signal analysis.

Shock calibration requires calibration of all the elements used in the measurement system. Calibration of the piezo-accelerometer is carried out using a shock machine of the form shown in Figure 15.2. The accelerometer is mounted, together with a standard accelerometer, on the anvil of the machine, and records of the shock signals from the two transducers are recorded on a transient recorder and compared. If optical sensors are used to measure the time Δt taken for the anvil to move a fixed distance Δx, an absolute measurement of acceleration can also be obtained according to the following equation:

$$A = \int_0^{\Delta t} a(t) \, dt = \Delta V = \frac{\Delta x}{\Delta t}$$

where $a(t)$ is the instantaneous acceleration at time t and ΔV is the change in velocity of the anvil as it moves through Δx.

Traceability to fundamental standards is then provided by the procedures described in the last section on vibration measurement.

15.2 Viscosity

Viscosity measurement is important in many process industries. In the food industry, the viscosity of raw materials such as dough, batter and ice cream has a direct effect on the quality of the product. Similarly, in other industries such as ceramics, the quality of raw materials affects the final product quality. Viscosity control is also very important in assembly operations which involve the application of mastics and glue flowing through tubes. Clearly, successful assembly requires such materials to flow through tubes at the correct rate and therefore it is essential that their viscosity is correct.

Viscosity describes the way in which a fluid flows when it is subject to an applied force. Consider an elemental cubic volume of fluid and a shear force F applied to one of its faces of area A. If this face moves a distance L and at a velocity V relative to the opposite face of the cube under the action of F, the shear stress (τ) and shear rate (γ) are given by

$$\tau = \frac{F}{A} \quad ; \quad \gamma = \frac{V}{L}$$

The coefficient of viscosity (η) is the ratio of shear stress to shear rate, i.e.

$$\eta = \frac{\tau}{\gamma}$$

η is often decribed simply as the 'viscosity'. A further term, kinematic viscosity,

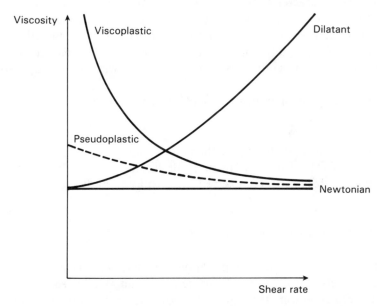

Figure 15.3 Different viscosity/shear rate relationships

is also sometimes used, given by

$$\nu = \frac{\eta}{\rho}$$

where ν is the kinematic viscosity and ρ is the fluid density. To avoid confusion, η is often known as the dynamic viscosity, to distinguish it from ν. η is measured in units of poise or Ns/m^2 and ν is measured in units of Stokes or m^2/s.

Viscosity was originally defined by Newton, who assumed that it was constant with respect to shear rate. However, it has since been shown that the viscosity of many fluids varies significantly at high shear rates and the viscosity of some varies even at low shear rates. The worst non-Newtonian characteristics tend to occur with emulsions, pastes and slurries. For non-Newtonian fluids, a sub-division into further classes can also be made according to the manner in which the viscosity varies with shear rate, as shown in Figure 15.3.

15.2.1 Viscosity measurement

The relationship between the input variables and output measurement for instruments which measure viscosity normally assumes that the measured fluid has Newtonian characteristics. For non-Newtonian fluids, a correction must be made for shear rate variations.[1] If such a correction is not made, the measurement obtained is known as the *apparent viscosity* and this can differ from the true viscosity by a large factor. The true viscosity is often called the *absolute viscosity* to avoid ambiguity. Viscosity also varies with fluid temperature and density.

Instruments for measuring viscosity work on one of three physical principles:

  (a) rate of flow of the liquid through a tube;
  (b) rate of fall of a body through the liquid;
  (c) viscous friction force exerted on a rotating body.

(i) Capillary and tube viscometers

These are the most accurate type of viscometer, with typical measurement inaccuracy levels down to ± 0.3 per cent. Liquid is allowed to flow, under gravity from a reservoir, through a tube of known cross-section. In different instruments, the tube can vary from capillary-sized to a large diameter. The

pressure difference across the ends of the tube and the time for a given quantity
of liquid to flow are measured, and then the liquid viscosity for Newtonian
fluids can be calculated as (in units of poise)

$$\eta = \frac{1.25\pi R^4 PT}{LV} \tag{15.2}$$

where R is the radius (m) of the tube, L is its length (m), P is the pressure difference (N/m^2) across the ends and V is the volume of liquid flowing in time T
(m^3/s). For non-Newtonian fluids, corrections must be made for shear rate variations.[1]

For any given viscometer, R, L and V are constant and equation (15.2) can
be written as

$$\eta = KPT \tag{15.3}$$

where K is known as the *viscometer constant*.

(ii) Falling body viscometer

The falling body viscometer is particularly recommended for the measurement
of high-viscosity fluids and can give measurement uncertainty levels down to
± 1 per cent. It involves measuring the time taken for a spherical body to fall
a given distance through the liquid. The viscosity for Newtonian fluids is then
given by Stoke's formula as (in units of poise)

$$\eta = \frac{R^2 g(\rho_s - \rho_l)}{450V}$$

where R is the radius (m) of the sphere, g is the acceleration due to gravity
(m/s^2), ρ_s and ρ_l are the specific gravities (g/m^3) of the sphere and liquid,
respectively, and V is the velocity (m/s) of the sphere. For non-Newtonian
fluids, correction for the variation in shear rate is very difficult.

(iii) Rotational viscometers

Rotational viscometers are relatively easy to use but their measurement inaccuracy is at least ± 10 per cent. All types have some form of element rotating
inside the liquid at a constant rate. One common version has two coaxial cylinders with the fluid to be measured contained between them. One cylinder is

driven at a constant angular velocity by a motor and the other is suspended by torsion wire. After the driven cylinder starts from rest, the suspended cylinder rotates until an equilibrium position is reached where the force due to the torsion wire is just balanced by the viscous force transmitted through the liquid. The viscosity (in poise) for Newtonian fluids is then given by

$$\eta = 2.5G \, \frac{\left(\dfrac{1}{R_1^2} - \dfrac{1}{R_2^2} \right)}{\pi h \omega}$$

where G is the couple (N·m) formed by the force exerted by the torsion wire and its deflection, R_1 and R_2 are the radii (m) of the inner and outer cylinders, h is the length of the cylinder (m) and ω is the angular velocity (rad/s) of the rotating cylinder. Again, corrections have to be made for non-Newtonian fluids.

15.2.2 Viscosity calibration

Viscosity standards are only defined for Newtonian fluids. No such standards exist for non-Newtonian fluids.

Viscosity calibration for Newtonian liquids is carried out using a set of glass capillary tubes with accurately known viscometer constants. These provide reference standards over a range of viscosities. The set is calibrated by a 'step-up' procedure from a calibration on distilled water, using a series of standard liquids. The steps in this procedure are as follows:

A 'master' viscometer is calibrated first using distilled water as the flowing fluid. The 'master' is then used to measure the viscosity of the first standard viscosity liquid. This liquid is then used to calibrate the first standard viscometer. After this, the viscosity of the second standard liquid is measured by the first standard viscometer and then the second standard viscometer is calibrated using the second standard liquid. This procedure is repeated to calibrate the other standard viscometers in the set using the other standard liquids. A more thorough explanation of this procedure and a discussion of the types of standard liquid used can be found in reference 2.

It is common practice to maintain two sets of such standard viscometers, one as a working set for routine calibrations and one as a reference set for calibration checks on the working set. Higher-level calibration is provided by comparison with a set of standard glass-capillary viscometers maintained by national standards organizations. Such organizations also commonly supply a range of standard viscosity liquids as an alternative calibration source. Where standard liquids are kept for calibration, they must be checked periodically to ensure their continuing stability as they age.

As there are no absolute physical quantities that the calibration of viscosity values can be checked against, the various national standards organizations carry out cross-checks between each other's laboratories. These checks take two forms. One is to transport a reference set of viscometers between laboratories and to compare the viscometer constants measured, with appropriate correction for differences in the value of g (acceleration due to gravity). The other form of check is to compare measurements made on standard liquids at a specified temperature. Precise guidelines for these calibration procedures are given in references 3 and 4.

In the case of non-Newtonian fluids, as has already been stated, there are no formal viscosity standards. Therefore, the only course of action available for checking viscometers intended to measure non-Newtonian fluids is to calibrate them against standard-viscosity Newtonian fluids.

15.3 Moisture

There are many quality-related industrial requirements for the measurement of the moisture content of both solids, liquids and gases. The physical properties and storage stability of most solid materials is affected by their water content. There is also a statutory requirement to limit the moisture content in the case of many materials sold by weight. In consequence, the requirement for moisture measurement pervades a large number of industries involved in the manufacture of foodstuffs, pharmaceuticals, cement, plastics, textiles and paper.

Measurement of the water content in liquids is commonly needed for fiscal purposes, but is also often necessary to satisfy statutory requirements. The petrochemical industry has wide-ranging needs for moisture measurement in oil, etc. The food industry also needs to measure the water content of products such as beer and milk.

In the case of moisture in gases, the most common measurement is the amount of moisture in air. This is usually known as the humidity level. Humidity measurement and control is an essential requirement in many buildings, greenhouses and vehicles.

As there are several ways in which humidity can be defined, three separate terms, as follows, have evolved so that ambiguity can be avoided:

1. *Absolute humidity* is the mass of water in a unit volume of moist air.
2. *Specific humidity* is the mass of water in a unit mass of moist air.

3. *Relative humidity* is the ratio of the actual water vapour pressure in air to the saturation vapour pressure, usually expressed as a percentage.

15.3.1 Industrial moisture measurement techniques

Industrial methods for measuring moisture are based on the variation of some physical property of the material with moisture content. Many different properties can be used and thus the range of available techniques, as listed below, is large.

Electrical methods

Measuring the amount of absorption of *microwave energy* beamed through the material is the most common technique for measuring moisture content and is described in detail in references 5 and 6. Microwaves at wavelengths between 1 mm and 1 m are absorbed to a much greater extent by water than most other materials. Wavelengths of 30 mm or 100 mm are commonly used because 'off-the-shelf' equipment to produce these is readily available from instrument suppliers. The technique is suitable for moisture measurement in solids, liquids and gases at moisture-content levels up to 45 per cent, and measurement uncertainties down to ±0.3 per cent are possible.

The *capacitance moisture meter* uses the principle that the dielectric constant of materials varies according to their water content. Capacitance measurement is therefore related to moisture content. The instrument is useful for measuring moisture-content levels up to 30 per cent in both solids and liquids, and measurement uncertainty down to ±0.3 per cent has been claimed for the technique.[7] Drawbacks of the technique include (a) limited measurement resolution owing to the difficulty in measuring small changes in a relatively large standing capacitance value; and (b) difficulties when the sample has a high electrical conductivity. An alternative capacitance charge transfer technique has been reported which overcomes these problems by measuring the charge-carrying capacity of the material.[8] In this technique, wet and dry samples of the material are charged to a fixed voltage and then simultaneously discharged into charge-measuring circuits.

The *electrical conductivity* of most materials varies with moisture content and this therefore provides another means of measurement. It is cheap and can measure moisture levels up to 25 per cent. However, the presence of other conductive substances in the material, such as salts or acids, affects the measurement.

A further technique is to measure the frequency change in a *quartz crystal* which occurs as it absorbs moisture.

Neutron moderation

Neutron moderation measures moisture content using a radioactive source and a neutron counter. Fast neutrons emitted from the source are slowed down by hydrogen nuclei in the water, forming a cloud whose density is related to the moisture content. Measurements take a long time because the output density reading may take up to a few minutes to reach steady-state, according to the nature of the materials involved. Also, the method cannot be used with any materials containing hydrogen molecules, such as oils and fats, as these also slow down neutrons. Specific humidities up to 15 per cent (± 1 per cent error) can be measured.

Low-resolution nuclear magnetic resonance (NMR)

Low-resolution nuclear magnetic resonance involves subjecting the sample to both a uni-directional and an alternating radio-frequency (RF) magnetic field. The amplitude of the uni-directional field is varied cyclically, which causes resonance once per cycle in the coil producing the RF field. Under resonance conditions, protons are released from the hydrogen content of the water in the sample. These protons cause a measurable moderation of the amplitude of the RF oscillator waveform which is related to the moisture content of the sample. A fuller description of the technique can be found in reference 9.

Materials having their own natural hydrogen content cannot normally be measured. However, pulsed NMR techniques have been developed which overcome this problem by taking advantage of the different relaxation times of hydrogen nuclei in water and oil. In such pulsed techniques, the dependence on the relaxation time limits the maximum fluid flow rate for which moisture can be measured.

Optical methods

The *refractometer* is a well-established instrument which is used for measuring the water content of liquids. It measures the refractive index of the liquid which changes according to the moisture content.

Moisture-related *energy absorption* of near-infra-red light can be used for measuring the moisture content of solids, liquids and gases. At a wavelength of

1.94 μm, energy absorption due to moisture is high, whereas at 1.7 μm, absorption due to moisture is zero. Therefore, measuring absorption at both 1.94 μm and 1.7 μm allows absorption due to components in the material other than water to be compensated for, and the resulting measurement is directly related to energy content. The latest instruments use multiple-frequency infra-red energy and have an even greater capability for eliminating the effect of components in the material other than water which absorb energy. Such multi-frequency instruments also cope much better with variations in particle size in the measured material.

In alternative versions of this technique, energy is either transmitted through the material or reflected from its surface. In either case, materials which are either very dark or highly reflective give poor results. The technique is particularly attractive, where applicable, because it is a non-contact method which can be used to monitor moisture content continuously at moisture levels up to 50 per cent, with inaccuracy as low as ± 0.1 per cent in the measured moisture level. A deeper treatment can be found in reference 10.

Ultrasonic methods

The presence of water changes the speed of propagation of ultrasonic waves through liquids. The moisture content of liquids can therefore be determined by measuring the transmission speed of ultrasound. This has the inherent advantage of being a non-invasive technique but temperature compensation is essential because the velocity of ultrasound is particularly affected by temperature changes. The method is best suited to measurement of high moisture levels in liquids which are not aerated or of high viscosity. Typical measurement uncertainty is ± 1 per cent but measurement resolution is very high, with changes in moisture level as small as 0.05 per cent being detectable. Further details can be found in reference 11.

Mechanical properties

Density changes in many liquids and slurries can be measured and related to moisture content, with good measurement resolution up to 0.2 per cent moisture. Moisture content can also be estimated by measuring the moisture-level-dependent viscosity of liquids, pastes and slurries.

15.3.2 Laboratory techniques for moisture measurement

Laboratory techniques for measuring moisture content generally take much longer to obtain a measurement than the industrial techniques described above. However, the measurement accuracy obtained is usually much better.

Water separation

Various laboratory techniques are available which enable the moisture content of liquids to be measured accurately by separating the water from a sample of the host liquid. Separation is effected by either titration (Karl Fischer technique), distillation (Dean and Stark technique) or a centrifuge. Any of these methods can measure water content in a liquid with measurement uncertainty levels down to ± 0.03 per cent.

Gravimetric methods

Moisture content in solids can be measured accurately by weighing the moist sample, drying it and then weighing again. Great care must be taken in applying this procedure, as many samples rapidly take up moisture again if they are removed from the drier and exposed to the atmosphere before being weighed. Normal procedure is to put the sample in an open container, dry it in an oven and then screw an airtight top onto the container before it is removed from the oven.

Phase-change methods

The boiling and freezing point of materials is altered by the presence of moisture, and therefore the moisture level can be determined by measuring the phase-change temperature. This technique is used for measuring the moisture content in many food products and in some oil and alcohol products.

Equilibrium relative humidity measurement

This technique involves placing a humidity sensor in close proximity to the sample in an airtight container. The water vapour pressure close to the sample is related to the moisture content of the sample. The moisture level can therefore be determined from the humidity measurement.

15.3.3 Humidity measurement

The three major instruments used for measuring humidity in industry are the electrical hygrometer, the psychrometer and the dew point meter. The dew point meter, described in Section 15.3.4, is the most accurate of these and is used as a calibration standard. A more detailed description of these and other types of hygrometer can be found in reference 12.

The electrical hygrometer

The electrical hygrometer measures the change in capacitance or conductivity of a hygroscopic material as its moisture level changes. Conductivity types use two noble-metal electrodes on either side of an insulator coated in a hygroscopic salt such as calcium chloride. Capacitance types have two plates on either side of a hygroscopic dielectric such as aluminium oxide.

These instruments are suitable for measuring moisture levels between 15 per cent and 95 per cent, with typical measurement uncertainty of ±3 per cent. Atmospheric contaminants and operation in saturation conditions both cause characteristics drift, and therefore the re-calibration frequency has to be determined according to the conditions of use.

The psychrometer (wet and dry bulb hygrometer)

The psychrometer, also known as the wet and dry bulb hygrometer, has two temperature sensors, one exposed to the atmosphere and one enclosed in a wet wick. Air is blown across the sensors which causes evaporation and a reduction in temperature in the wet sensor. The temperature difference between the sensors is related to the humidity level. The lowest measurement uncertainty attainable is ±4 per cent.

15.3.4 Calibration

Microwave techniques and instruments such as the electrical hygrometer and dew point meter are accurate enough for working standard calibrations. Otherwise, such calibrations are performed by gravimetric or water-separation laboratory techniques (see Section 15.3.2). For calibration to secondary and primary reference standards, test samples must be used in which the water content is accurately known.

Electrical hygrometer

The electrical hygrometer, as described in the last section, can be used for first-level calibrations of instruments in regular use provided that it is not subjected to near-saturation conditions and is stored carefully away from possible contaminants.

Microwave techniques

Measuring the absorption of microwaves, as described in Section 15.3.1, is also accurate enough to be used as a first-level calibration technique for instruments in everyday use, as described in references 5 and 6.

Dew point meter

The elements of the dew point meter, also known as the dew point hygrometer, are shown in Figure 15.4. The sample is introduced into a vessel with an electrically cooled mirror surface. The mirror surface is cooled until a light source–light detector system detects the formation of dew on the mirror, and

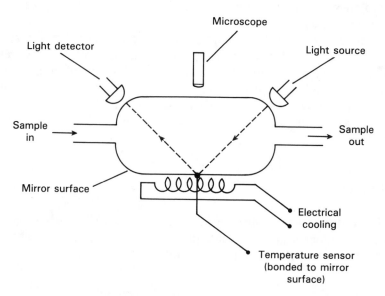

Figure 15.4 Dew point meter

the condensation temperature is measured by a sensor bonded to the mirror surface. The dew point is the temperature at which the sample becomes saturated with water. Therefore, this temperature is related to the moisture level in the sample. A microscope is also provided in the instrument so that the thickness and nature of the condensate can be observed. The instrument is described in greater detail in reference 13.

Even small levels of contaminants on the mirror surface can cause large changes in the dew point and therefore the instrument must be kept very clean. When necessary, the mirror should be cleaned with deionized or distilled water applied with a lint-free swab. Any contamination can be detected by a skilled operator, as this makes the condensate look 'blotchy' when viewed through the microscope. The microscope also shows up other potential problems such as large ice crystals in the condensate which cause temperature gradients between the condensate and the temperature sensor.

At suitable intervals, the instrument must itself be calibrated against a test gas containing a known amount of water.

Test samples

The traceability of moisture measurements to reference standards is achieved with the provision by national standards organizations of test samples which have an accurately known water content. The instrument to be calibrated is applied to these samples to see whether its output reading agrees with the known water content.

Test samples created in a company's own calibration laboratory can be used as working standards which are checked against primary reference standards at suitable intervals. This involves making up mixtures of a gas or liquid with water in known proportions. In the case of solids, it is often difficult to make up a test sample with an even distribution of water through it, and gravimetric techniques of calibration (see Section 15.3.2) have to be used instead.

15.4 Volume

Volume measurement is required in its own right as well as being required as a necessary component in some techniques for the measurement and calibration of other quantities such as volume flow rate and viscosity.

15.4.1 Volume measurement

The volume of vessels of a regular shape, where the cross-section is circular or oblong in shape, can be readily calculated from the dimensions of the vessel, using instruments as described in Chapter 13. Otherwise, for vessels of irregular shape, either gravimetric techniques or a set of calibrated volumetric measures are required.

In the gravimetric technique, the dry vessel is weighed and then completely filled with water and weighed again. The alternative technique involves transferring the liquid from the vessel into an appropriate number of volumetric measures taken from a standard-capacity, calibrated set. Each vessel in the set has a mark which shows the volume of liquid contained when the vessel is filled up to the mark. Special care is needed to ensure that the meniscus of the water is in the correct position with respect to the reference mark on the vessel when it is deemed to be full. Normal practice is to set the water level such that the reference mark forms a smooth tangent with the convex side of the meniscus. This is made easier to achieve if the meniscus is viewed against a white background and the vessel is shaded from stray illumination.

15.4.2 Volume calibration

Either of the latter two techniques may be used for calibration purposes, although the gravimetric technique gives better accuracy.[14]

Gravimetric technique

When the gravimetric technique, as described above, is used for volumetric calibration, certain special procedures should be followed. Firstly, the water used to fill the vessel should be either distilled or deionized. (Where very large vessels are being calibrated and it is impractical to use other than tap water, the difference in water density compared with pure water gives a typical error of 0.02 per cent.) Secondly, the calibration should be carried out under specified temperature and pressure conditions, and sufficient time should be allowed for the vessel and water to reach close thermal equilibrium. This latter condition is most easily achieved by maintaining a store of distilled or deionized water in the calibration laboratory, which ensures that the water is automatically at the correct temperature. Lastly, correction may have to be made for changes in air buoyancy, which become significant if calibration is carried out at altitudes greater than 150 m.

Calibrated measures

The measurement uncertainty using calibrated volumetric measures depends on the number of measures used for any particular measurement. The total error is a multiple of the individual error of each measure, typical values of which are shown in Table 15.1.

As mentioned in the last section, special care must be taken to ensure that the menicus level is set correctly with respect to the reference mark when each vessel is filled. Calibration traceability is provided by calibrating all volumetric measures gravimetrically at suitable intervals of time.

Table 15.1 Typical measurement uncertainty of volumetric measures

Capacity	Volumetric uncertainty (%)
1 ml	±4
10 ml	±0.8
100 ml	±0.2
1 l	±0.1
10 l	±0.05
100 l	±0.02
1000 l	±0.02

Appendix 1
Fundamental and derived SI units

(a) Fundamental units

Quantity	Standard unit	Symbol
length	metre	m
mass	kilogram	kg
time	second	s
electric current	ampere	A
temperature	degrees Kelvin	K
luminous intensity	candela	cd
matter	mole	mol

(b) Supplementary fundamental units

Quantity	Standard unit	Symbol
plane angle	radian	rad
solid angle	steradian	sr

(c) Derived units

Quantity	Standard unit	Symbol	Derivation formula
area	square metre	m^2	
volume	cubic metre	m^3	
velocity	metre per second	m/s	
acceleration	metre per second squared	m/s^2	
angular velocity	radian per second	rad/s	
angular acceleration	radian per second squared	rad/s^2	
density	kilogram per cubic metre	kg/m^3	
specific volume	cubic metre per kilogram	m^3/kg	
mass flow rate	kilogram per second	kg/s	
volume flow rate	cubic metre per second	m^3/s	
force	newton	N	$kg \cdot m/s^2$
pressure	newton per square metre	N/m^2	
torque	newton-metre	$N \cdot m$	
momentum	kilogram-metre per second	$kg \cdot m/s$	
moment of inertia	kilogram-metre squared	$kg \cdot m^2$	
kinematic viscosity	square metre per second	m^2/s	
dynamic viscosity	newton-second per sq metre	$N \cdot s/m^2$	
work, energy, heat	joule	J	$N \cdot m$
specific energy	joule per cubic metre	J/m^3	
power	watt	W	J/s
thermal conductivity	watt per metre-Kelvin	W/m K	
electric charge	coulomb	C	$A \cdot s$
voltage, emf, pot diff	volt	V	W/A
electric field strength	volt per metre	V/m	
electric resistance	ohm	Ω	V/A
electric capacitance	farad	F	$A \cdot s/V$
electric inductance	henry	H	$V \cdot s/A$
electric conductance	siemen	S	A/V
resistivity	ohm-metre	$Ω \cdot m$	
permittivity	farad per metre	F/m	
permeability	henry per metre	H/m	
current density	ampere per square metre	A/m^2	
magnetic flux	weber	Wb	$V \cdot s$
magnetic flux density	tesla	T	Wb/m^2
magnetic field strength	ampere per metre	A/m	
frequency	hertz	Hz	s^{-1}
luminous flux	lumen	lm	$cd \cdot sr$
luminance	candela per square metre	cd/m^2	
illumination	lux	lx	lm/m^2
molar volume	cubic metre per mole	m^3/mol	
molarity	mole per kilogram	mol/kg	
molar energy	Joule per mole	J/mol	

Appendix 2

Imperial–metric–SI conversion tables

Length

SI units: mm, m, km
Imperial units: in, ft, mile

	mm	m	km	in	ft	mile
mm	1	10^{-3}	10^{-6}	0.0393701	3.281×10^{-3}	–
m	1000	1	10^{-3}	39.3701	3.28084	6.214×10^{-4}
km	10^6	10^3	1	39370.1	3280.84	0.621371
in	25.4	0.0254	–	1	0.083333	–
ft	304.8	0.3048	3.048×10^{-4}	12	1	1.894×10^{-4}
mile	–	1609.34	1.60934	63360	5280	1

Area

SI units: mm^2, m^2, km^2
Imperial units: in^2, ft^2, $mile^2$

	mm^2	m^2	km^2	in^2	ft^2	$mile^2$
mm^2	1	10^{-6}	–	1.550×10^{-3}	1.076×10^{-5}	–
m^2	10^6	1	10^{-6}	1550	10.764	–
km^2	–	10^6	1	–	1076×10^7	0.3861
in^2	645.16	6.452×10^{-4}	–	1	6.944×10^{-3}	–
ft^2	92903	0.09290	–	144	1	–
$mile^2$	–	2.590×10^6	2.590	–	2.788×10^7	1

Second moment of area

SI units: mm^4, m^4
Imperial units: in^4, ft^4

	mm^4	m^4	in^4	ft^4
mm^4	1	10^{-12}	2.4025×10^{-6}	1.159×10^{-10}
m^4	10^{12}	1	2.4025×10^{6}	115.86
in^4	416231	4.1623×10^{-7}	1	4.8225×10^{-5}
ft^4	8.631×10^{9}	8.631×10^{-3}	20736	1

Volume

SI units: mm^3, m^3
Metric units: ml, l
Imperial units: in^3, ft^3, UK gallon

	mm^3	ml	l	m^3	in^3	ft^3	UK gallon
mm^3	1	10^{-3}	10^{-6}	10^{-9}	6.10×10^{-5}	–	–
ml	10^3	1	10^{-3}	10^{-6}	0.061024	3.53×10^{-5}	2.2×10^{-4}
l	10^6	10^3	1	10^{-3}	61.024	0.03532	0.22
m^3	10^9	10^6	10^3	1	61024	35.31	220
in^3	16387	16.39	0.0164	1.64×10^{-5}	1	5.79×10^{-4}	3.61×10^{-3}
ft^3	–	2.83×10^{4}	28.32	0.02832	1728	1	6.229
UK gall	–	4546	4.546	4.55×10^{-3}	277.4	0.1605	1

Note:
Additional unit: 1 US gallon = 0.8327 UK gallon.

Density

SI unit: kg/m^3
Metric unit: g/cm^3
Imperial units: lb/ft^3, lb/in^3

	kg/m^3	g/cm^3	lb/ft^3	lb/in^3
kg/m^3	1	10^{-3}	0.062428	3.605×10^{-5}
g/cm^3	1000	1	62.428	0.036127
lb/ft^3	16.019	0.016019	1	5.787×10^{-4}
lb/in^3	27680	27.680	1728	1

Mass

SI units: g, kg, t
Imperial units: lb, cwt, ton

	g	kg	t	lb	cwt	ton
g	1	10^{-3}	10^{-6}	2.205×10^{-3}	1.968×10^{-5}	9.842×10^{-7}
kg	10^3	1	10^{-3}	2.20462	0.019684	9.842×10^{-4}
t	10^6	10^3	1	2204.62	19.6841	0.984207
lb	453.592	0.45359	4.536×10^{-4}	1	8.929×10^{-3}	4.464×10^{-4}
cwt	50802.3	50.8023	0.050802	112	1	0.05
ton	1.016×10^6	1016.05	1.01605	2240	20	1

Force

SI units: N, kN
Metric unit: kg_f
Imperial units: pdl (poundal), lb_f, UK ton_f

	N	kg_f	kN	pdl	lb_f	UK ton_f
N	1	0.1020	10^{-3}	7.233	0.2248	1.004×10^{-4}
kg_f	9.807	1	9.807×10^{-3}	70.93	2.2046	9.842×10^{-4}
kN	1000	102.0	1	7233	224.8	0.1004
pdl	0.1383	0.0141	1.383×10^{-4}	1	0.0311	1.388×10^{-5}
lb_f	4.448	0.4536	4.448×10^{-3}	32.174	1	4.464×10^{-4}
UK ton_f	9964	1016	9.964	72070	2240	1

Note:
Additional unit: 1 dyne $= 10^{-5}$N $= 7.233 \times 10^{-5}$ pdl.

Torque (moment of force)

SI unit: Nm
Metric unit: $kg_f m$
Imperial units: pdl ft, lb_fft

	Nm	$kg_f m$	pdl ft	lb_fft
Nm	1	0.1020	23.73	0.7376
$kg_f m$	9.807	1	232.7	7.233
pdl ft	0.04214	4.297×10^{-3}	1	0.03108
lb_fft	1.356	0.1383	32.17	1

Inertia

SI unit: Nm^2
Imperial unit: $lb_f ft^2$

$$1\ lb_f ft^2 = 0.4132\ Nm^2$$
$$1\ Nm^2 = 2.420\ lb_f ft^2$$

Pressure

SI units: mbar, bar, N/m^2
Imperial units: lb/in^2, in Hg, atmosphere

	mbar	bar	N/m^2	lb/in^2	in Hg	atmosphere
mbar	1	10^{-3}	100	0.01450	0.02953	9.869×10^{-4}
bar	1000	1	10^5	14.50	29.53	0.9869
N/m^2	0.01	10^{-5}	1	1.450×10^{-4}	2.953×10^{-4}	9.869×10^{-6}
lb/in^2	68.95	0.06895	6895	1	2.036	0.06805
in Hg	33.86	0.03386	3386	0.4912	1	0.03342
atmos	1013	1.013	1.013×10^5	14.70	29.92	1

Additional conversion factors

1 inch water $= 0.073\ 56$ inch Hg $= 2.491$ mbar
1 torr $= 1.333$ mbar
1 Pascal $= 1\ N/m^2$

Energy, work, heat

SI unit: J
Metric units: $kg_f \cdot m$, $kW \cdot h$
Imperial units: $ft \cdot lb_f$, cal, Btu

	J	$kg_f \cdot m$	$kW \cdot h$	$ft \cdot lb_f$	cal	Btu
J	1	0.1020	2.778×10^{-7}	0.7376	0.2388	9.478×10^{-4}
$kg_f \cdot m$	9.8066	1	2.724×10^{-6}	7.233	2.342	9.294×10^{-3}
$kW \cdot h$	3.600×10^6	367098	1	2.655×10^6	859845	3412.1
$ft \cdot lb_f$	1.3558	0.1383	3.766×10^{-7}	1	0.3238	1.285×10^{-3}
cal	4.1868	0.4270	1.163×10^{-6}	3.0880	1	3.968×10^{-3}
Btu	1055.1	107.59	2.931×10^{-4}	778.17	252.00	1

Additional conversion factors

1 therm = 10^5 Btu = 1.0551×10^8 J.
1 thermie = 4.186×10^6 J.
1 hp·h = 0.7457 kW·h = $2.6845 \cdot 10^6$ J.
1 ft·pdl = 0.042 14 J.
1 erg = 10^{-7} J.

Power

SI units: W, kW
Imperial units: HP, ft·lb$_f$/s

	W	kW	HP	ft·lb$_f$/s
W	1	10^{-3}	1.341×10^{-3}	0.73564
kW	10^3	1	1.34102	735.64
HP	745.7	0.7457	1	548.57
ft·lb$_f$/s	1.35935	1.359×10^{-3}	1.823×10^{-3}	1

Velocity

SI units: mm/s, m/s
Metric unit: km/h
Imperial units: ft/s, mile/h

	mm/s	m/s	km/h	ft/s	mile/h
mm/s	1	10^{-3}	3.6×10^{-3}	3.281×10^{-3}	2.237×10^{-3}
m/s	1000	1	3.6	3.28084	2.23694
km/h	277.778	0.277778	1	0.911344	0.621371
ft/s	304.8	0.3048	1.09728	1	0.681818
mile/h	447.04	0.44704	1.609344	1.46667	1

Acceleration

SI units: m/s^2
Other metric unit: cm/s^2
Imperial unit: ft/s^2
Other unit: g

	m/s^2	cm/s^2	ft/s^2	g
m/s^2	1	100	3.281	0.102
cm/s^2	0.01	1	0.0328	0.00102
ft/s^2	0.3048	30.48	1	0.03109
g	9.81	981	32.2	1

Mass flow rate

SI unit: g/s
Metric units: kg/h, tonne/d
Imperial units: lb/s, lb/h, ton/d

	g/s	kg/h	tonne/d	lb/s	lb/h	ton/d
g/s	1	3.6	0.08640	2.205×10^{-3}	7.937	0.08503
kg/h	0.2778	1	0.02400	6.124×10^{-4}	2.205	0.02362
tonne/d	11.57	41.67	1	0.02551	91.86	0.9842
lb/s	453.6	1633	39.19	1	3600	38.57
lb/h	0.1260	0.4536	0.01089	2.788×10^{-4}	1	0.01071
ton/d	11.76	42.34	1.016	0.02593	93.33	1

Volume flow rate

SI unit: m^3/s
Metric units: l/h, ml/s
Imperial units: gal/h, ft^3/s, ft^3/h

	l/h	ml/s	m^3/s	gal/h	ft^3/s	ft^3/h
l/h	1	0.2778	2.778×10^{-7}	0.2200	9.810×10^{-6}	0.035316
ml/s	3.6	1	10^{-6}	0.7919	3.532×10^{-5}	0.12714
m^3/s	3.6×10^6	10^6	1	7.919×10^5	35.31	1.271×10^5
gal/h	4.546	1.263	1.263×10^{-6}	1	4.460×10^{-5}	0.16056
ft^3/s	1.019×10^5	2.832×10^4	0.02832	2.242×10^4	1	3600
ft^3/h	28.316	7.8653	7.865×10^{-6}	6.2282	2.778×10^{-4}	1

Specific energy (heat per unit volume)

SI units: J/m^3, kJ/m^3, MJ/m^3
Imperial units: $kcal/m^3$, Btu/ft^3, therm/UK gal

	J/m^3	kJ/m^3	MJ/m^3	$kcal/m^3$	Btu/ft^3	therm/UK gal
J/m^3	1	10^{-3}	10^{-6}	1.388×10^{-4}	2.684×10^{-5}	–
kJ/m^3	1000	1	10^{-3}	0.2388	0.02684	–
MJ/m^3	10^6	1000	1	238.8	26.84	4.309×10^{-5}
$kcal/m^3$	4187	4.187	4.187×10^{-3}	1	0.1124	1.804×10^{-7}
Btu/ft^3	3.726×10^4	37.26	0.03726	8.899	1	1.605×10^{-6}
therm/gal	–	–	2.321×10^4	5.543×10^6	6.229×10^5	1

Dynamic viscosity

SI unit: Ns/m^2
Metric unit: cP (centipoise), P (poise) $[1\ P = 100\ g/m \cdot s]$
Imperial unit: $lb_m/ft \cdot h$

	$lb_m/ft \cdot h$	P	cP	Ns/m^2
$lb_m/ft \cdot h$	1	4.133×10^{-3}	0.4134	4.134×10^{-4}
P	241.9	1	100	0.1
cP	2.419	0.01	1	10^{-3}
Ns/m^2	2419	10	1000	1

Note:
Additional unit: 1 Pascal second = 1 Ns/m^2.

Kinematic viscosity

SI unit: m^2/s
Metric unit: cSt (centistokes), St (Stokes)
Imperial unit: ft^2/s

	ft^2/s	m^2/s	cSt	St
ft^2/s	1	0.0929	9.29×10^4	929
m^2/s	10.764	1	10^6	10^4
cSt	1.0764×10^{-5}	10^{-6}	1	0.01
St	1.0764×10^{-3}	10^{-4}	100	1

Appendix 3
Dynamic characteristics of instruments

The dynamic characteristics of a measuring instrument describe its behaviour between the time a measured quantity changes value and the time when the instrument output response attains a steady value.

These characteristics are formally defined mathematically as follows: In any linear, time-invariant measuring system, the following general relation can be written between the instrument's input and output for time $(t) > 0$:

$$a_n \frac{d^n q_o}{dt^n} + a_{n-1} \frac{d^{n-1} q_o}{dt^{n-1}} + \dots + a_1 \frac{dq_o}{dt} + a_o q_o$$

$$= b_m \frac{d^m q_i}{dt^m} + b_{m-1} \frac{d^{m-1} q_i}{dt^{m-1}} + \dots + b_1 \frac{dq_i}{dt} + b_o q_i \quad \text{(A3.1)}$$

where q_i is the measured quantity and q_o is the output reading. If we limit consideration to that of step changes in the measured quantity only, then equation (A3.1) reduces to

$$a_n \frac{d^n q_o}{dt^n} + a_{n-1} \frac{d^{n-1} q_o}{dt^{n-1}} + \dots + a_1 \frac{dq_o}{dt} + a_o q_o = b_o q_i \quad \text{(A3.2)}$$

Further simplification can be made by taking certain special cases of equation (A3.2) which collectively apply to nearly all measurement systems.

(a) Zero order instrument

If all the coefficients $a_1, \dots, a_n$ other than a_o in equation (A3.2) are assumed zero, then

$$a_o q_o = b_o q_i \qquad \text{or} \qquad q_o = \frac{b_o q_i}{a_o} = K q_i \quad \text{(A3.3)}$$

where K is a constant known as the instrument sensitivity as defined in Chapter 4.

Any instrument which behaves according to equation (A3.3) has negligible dynamic characteristics and is known as a zero order instrument.

278

(b) First order instrument

If all the coefficients $a_2, ..., a_n$ except for a_0 and a_1 are assumed zero in equation (A3.2) then

$$a_1 \frac{dq_o}{dt} + a_0 q_o = b_0 q_i \qquad (A3.4)$$

Any instrument which behaves according to equation (A3.4) is known as a first order instrument. If d/dt is replaced by the D operator in equation (A3.4), we obtain

$$a_1 D q_o + a_0 q_o = b_0 q_i$$

and rearranging:

$$q_o = \frac{(b_0/a_0)q_i}{(1 + \{a_1/a_0\}D)} \qquad (A3.5)$$

Defining $K = b_0/a_0$ as the static sensitivity and $\tau = a_1/a_0$ as the time constant of the system, equation (A3.5) becomes

$$q_o = \frac{Kq_i}{1 + \tau D} \qquad (A3.6)$$

If equation (A3.6) is solved analytically, the output quantity q_o in response to a step change in q_i varies with time in the manner shown in Figure 4.10. The time constant τ of the step response is the time taken for the output quantity q_o to reach 63 per cent of its final value.

Example
A balloon is equipped with temperature- and altitude-measuring instruments and has radio equipment which can transmit the output readings of these instruments back to ground. The balloon is initially anchored to the ground with the instrument output readings in steady-state. The altitude-measuring instrument is approximately zero order and the temperature transducer first order with a time constant of 15 s. The temperature on the ground, T_o, is 10 °C and the temperature T_x at an altitude of x metres is given by the relation

$$T_x = T_o - 0.01x$$

1. If the balloon is released at time zero, and thereafter rises upwards at a velocity of 5 m/s, draw a table showing the temperature and altitude measurements reported at intervals of 10 s over the first 50 s of travel. Show also in the table the error in each temperature reading.
2. What temperature does the balloon report at an altitude of 5000 m?

Solution
(Note − some knowledge of solving first order differential equations is assumed in the solution presented here. Readers without this knowledge will not be able to follow the solution easily but should not worry unduly. The main purpose of the exercise is to emphasize the physical effect of first order lags on measurements, and it is not necessary to fully understand the mathematical analysis in gaining this appreciation.)

Let the temperature reported by the balloon at some general time t be T_r. Then T_x is related to T_r by the relation

$$T_r = \frac{T_x}{1 + \tau D} = \frac{T_o - 0.01x}{1 + \tau D} = \frac{10 - 0.01x}{1 + 15D}$$

It is given that $x = 5t$. Thus

$$T_r = \frac{10 - 0.05t}{1 + 15D}$$

The complimentary function part of the solution is given by

$$T_{rcf} = C \exp(-t/15)$$

The particular integral part of the solution is given by

$$T_{rpi} = 10 - 0.05(t - 15)$$

Thus the whole solution is given by

$$T_r = T_{rcf} + T_{rpi} = C \exp(-t/15) + 10 - 0.05(t - 15)$$

Applying initial conditions:

At $t = 0$, $T_r = 10$

i.e.

$$10 = Ce^{-0} + 10 - 0.05(-15)$$

Thus $C = -0.75$ and the solution can be written as

$$T_r = 10 - 0.75 \exp(-t/15) - 0.05(t - 15)$$

(a) Using the above expression to calculate T_r for various values of t, the following table can be constructed:

Time	Altitude	Temp. reading	Temp. error
0	0	10	0
10	50	9.86	0.36
20	100	9.55	0.55
30	150	9.15	0.65
40	200	8.70	0.70
50	250	8.22	0.72

(b) At 5000 M, $t = 1000$ s. Calculating T_r from the above expression:

$$T_r = 10 - 0.75 \exp(-1000/15) - 0.05(1000 - 15)$$

The term in exp approximates to zero and so T_r can be written as

$$T_r \simeq 10 - 0.05(985) = -39.25\ ^{\circ}C$$

This result might have been inferred from the table above where it can be seen that the error is converging towards a value of 0.75. For large values of t, the transducer reading lags the true temperature value by a period of time equal to the time constant of 15 s. In this time, the balloon travels a distance of 75 m and the temperature falls by 0.75°. Thus for large values of t, the output reading is always 0.75° less than it should be.

(c) Second order instrument

If all coefficients $a_3, ..., a_n$ other than a_0, a_1 and a_2 in equation (A3.2) are assumed zero, then we obtain

$$a_2 \frac{d^2 q_o}{dt^2} + a_1 \frac{dq_o}{dt} + a_0 q_o = b_0 q_i \tag{A3.7}$$

Applying the D operator again:

$$a_2D^2q_o + a_1Dq_o + a_oq_o = b_oq_i$$

and re-arranging:

$$q_o = \frac{b_oq_i}{a_o + a_1D + a_2D^2} \tag{A3.8}$$

It is convenient to re-express the variables a_o, a_1, a_2 and b_o in equation (A3.8) in terms of three parameters K (static sensitivity), ω (undamped natural frequency) and ε (damping ratio), where

$$K = b_o/a_o$$

$$\omega = a_o/a_2$$

$$\varepsilon = \frac{a_1}{2a_oa_2}$$

Re-expressing equation (A3.8) in terms of K, ω and ε we obtain

$$\frac{q_o}{q_i} = \frac{K}{D^2/\omega^2 + 2\varepsilon D/\omega + 1} \tag{A3.9}$$

This is the standard equation for a second order system and any instrument whose response can be described by it is known as a second order instrument.

If equation (A3.9) is solved analytically, the shape of the step response obtained depends on the value of the damping ratio parameter ε. The output responses of a second order instrument for various values of ε are shown in Figure 4.11. For case (a) where $\varepsilon = 0$, there is no damping and the instrument output exhibits constant amplitude oscillations when disturbed by any change in the physical quantity measured. For light damping of $\varepsilon = 0.2$, represented by case (b), the response to a step change in input is still oscillatory but the oscillations gradually die down. Further increase in the value of ε reduces oscillations and overshoot still more, as shown by curves (c) and (d), and finally the response becomes very overdamped as shown by curve (e) where the output reading creeps up slowly towards the correct reading. Clearly, the extreme response curves (a) and (e) are grossly unsuitable for any measuring instrument. If an instrument were to be only ever subjected to step inputs, then the design strategy would be to aim towards a damping ratio of 0.707, which gives the critically damped response (c). Unfortunately, most of the physical quantities which instruments are required to measure do not change in the mathematically convenient form of steps, but rather in the form of ramps of varying slopes. As the form of the input variable changes, so the best value for ε varies, and choice of ε becomes one of compromise between those values that are best for each type of input variable behaviour anticipated. Commercial second order instruments, of which the accelerometer is a common example, are generally designed to have a damping ratio (ε) somewhere in the range 0.6–0.8.

Appendix 4
Measurement disturbance in electrical circuits – analysis by Thevenin's theorem

In analysing system disturbance during measurements in electric circuits, Thevenin's theorem is often of great assistance. Before going on to show how it is used, the principles of Thevenin's theorem will be explained first.

A4.1 Principles of Thevenin's theorem

Thevenin's theorem is extremely useful in the analysis of complex electrical circuits. It states that any network which has two accessible terminals A and B can be replaced, as far as its external behaviour is concerned, by a single e.m.f. acting in series with a single resistance between A and B. The single equivalent e.m.f. is that e.m.f. which is measured across A and B when the circuit external to the network is disconnected. The single equivalent resistance is the resistance of the network when all current and voltage sources within it are reduced to zero. To calculate this internal resistance of the network, all current sources within it are treated as open circuits and all voltage sources as short circuits. The proof of Thevenin's theorem can be found in the text by Skilling.[1]

Figure A4.1 shows part of a network consisting of a voltage source and four resistances. As far as its behaviour external to the terminals A and B are concerned, this can be regarded as a single voltage source V_t and a single resistance R_t. Applying Thevenin's theorem, R_t is found first of all by treating V_1

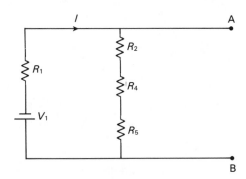

Figure A4.1 Example network 1

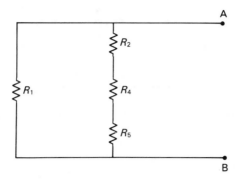

Figure A4.2 Network 1 with voltage source removed

as a short circuit, as shown in Figure A4.2. This is simply two resistances, R, and $(R_2 + R_4 + R_5)$ in parallel. The equivalent resistance R_t is thus given by

$$R_t = \frac{R_1(R_2 + R_4 + R_5)}{R_1 + R_2 + R_4 + R_5}$$

V_t is the voltage drop across AB. To calculate this, it is necessary to carry out an intermediate step of working out the current flowing, I. Referring to Figure A4.1, this is given by

$$I = \frac{V_1}{R_1 + R_2 + R_4 + R_5}$$

Now, V_t can be calculated from

$$V_t = I(R_2 + R_4 + R_5)$$

$$= \frac{V_1(R_2 + R_4 + R_5)}{R_1 + R_2 + R_4 + R_5}$$

The network of Figure A4.1 has thus been reduced to the simpler equivalent network shown in Figure A4.3.

Let us now proceed to the typical network problem of calculating the current flowing in the resistor R_3 of Figure A4.4. R_3 can be regarded as an external circuit or load on the rest of the network consisting of V_1, R_1, R_2, R_4 and R_5, as shown in Figure A4.5. This network of V_1, R_1, R_2, R_4 and R_5 is that shown in Figure A4.6. This can be rearranged to the network shown in Figure A4.1, which is equivalent to the single voltage source and resistance, V_t and R_t,

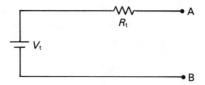

Figure A4.3 Equivalent circuit for network 1

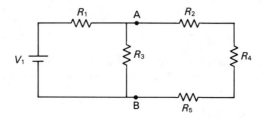

Figure A4.4 Example network 2

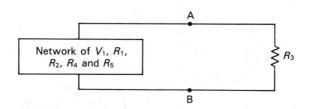

Figure A4.5 Alternative representation for network 2

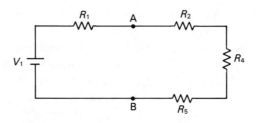

Figure A4.6 Network of V_1, R_1, R_2, R_4 and R_5

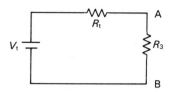

Figure A4.7 Equivalent circuit for network 2

calculated above. The whole circuit is then equivalent to that shown in Figure A4.7, and the current flowing through R_3 can be written as

$$I_{AB} = \frac{V_t}{R_t + R_3}$$

Thevenin's theorem can be applied successively to solve ladder networks of the form shown in Figure A4.8. Suppose in this network that it is required to calculate the current flowing in branch XY.

The first step is to imagine two terminals in the circuit, A and B, and regard the network to the right of AB as a load on the circuit to the left of AB. The circuit to the left of AB can be reduced to a single equivalent voltage source, E_{AB}, and resistance, R_{AB}, by Thevenin's theorem. If the 50 V source is replaced by its zero internal resistance (i.e. by a short circuit), then R_{AB} is given by

$$\frac{1}{R_{AB}} = \frac{1}{100} + \frac{1}{2000} = \frac{2000 + 100}{200\,000} \quad ; \quad \text{hence,} \; R_{AB} = 95.24\Omega$$

When AB is open-circuit, the current flowing round the loop to the left of AB is given by

$$I = \frac{50}{100 + 2000}$$

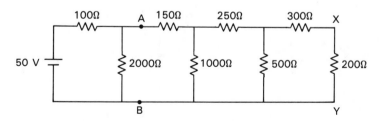

Figure A4.8 Example ladder network

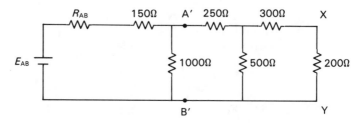

Figure A4.9 Simplification 1 for ladder network

Hence, E_{AB}, the open-circuit voltage across AB is given by

$$E_{AB} = I \cdot 2000 = 47.62 \text{ V}$$

We can now replace the circuit shown in Figure A4.8 by the simpler equivalent circuit shown in Figure A4.9.

The next stage is to apply an identical procedure to find an equivalent circuit consisting of voltage source, $E_{A'B'}$, and resistance, $R_{A'B'}$, for the network to the left of points A' and B' in Figure A4.9:

$$\frac{1}{R_{A'B'}} = \frac{1}{R_{AB} + 150} + \frac{1}{1000} = \frac{1}{245.24} + \frac{1}{1000} = \frac{1245.24}{245\,240}$$

Hence $R_{A'B'} = 196.94\Omega$

$$E_{A'B'} = \frac{1000}{R_{AB} + 150 + 1000} \cdot E_{AB} = 38.24 \text{ V}$$

The circuit can now be represented in the yet simpler form shown in Figure A4.10. Proceeding as before to find an equivalent voltage source and resistance,

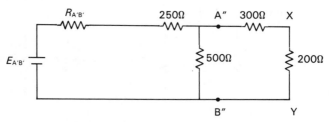

Figure A4.10 Simplification 2 for ladder network

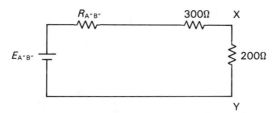

Figure A4.11 Final equivalent circuit for ladder network

$E_{A''B''}$ and $R_{A''B''}$, for the circuit to the left of A''B'' in Figure A4.10:

$$\frac{1}{R_{A''B''}} = \frac{1}{R_{A'B'} + 250} + \frac{1}{500} = \frac{500 + 446.94}{223\,470}$$

Hence $R_{A''B''} = 235.99\Omega$

$$E_{A''B''} = \frac{500}{R_{A'B'} + 250 + 500} \cdot E_{A'B'} = 20.19 \text{ V}$$

The circuit has how been reduced to the form shown in Figure A4.11, where the current through branch XY can be calculated simply as

$$I_{XY} = \frac{E_{A''B''}}{R_{A''B''} + 300 + 200} = \frac{20.19}{735.99} = 27.43 \text{ mA}$$

A4.2 Circuit loading during voltage measurement

Consider now the circuit shown in Figure A4.12(a), comprising two voltage sources and five resistors, in which the voltage across resistor R_5 is to be measured by a voltmeter with resistance R_m. Here, R_m acts as a shunt resistance across R_5, decreasing the resistance between points A and B and so disturbing the circuit. The voltage E_m measured by the meter is therefore not the value of the voltage E_o that existed prior to measurement. The extent of the disturbance can be assessed by calculating the open-circuit voltage E_o and comparing it with E_m.

Thevenin's theorem allows the circuit to be replaced by an equivalent circuit containing a single resistance and a single voltage source, as shown in Figure A4.12(b). For the purpose of defining the equivalent single resistance of a circuit by Thevenin's theorem, all voltage sources are represented by their internal resistance only, which can be approximated to zero, as shown in Figure A4.12(c). Analysis proceeds by calculating the equivalent resistances of sections

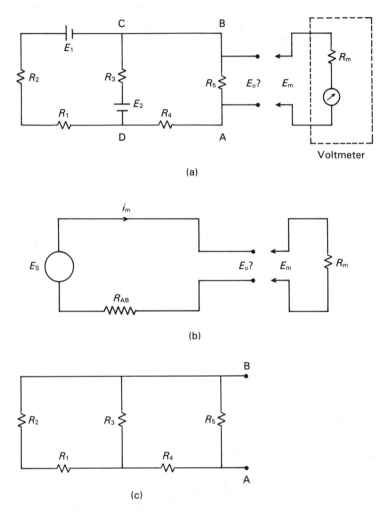

Figure A4.12: (a) a circuit in which the voltage across R_5 is to be measured; (b) equivalent circuit by Thevenin's theorem; (c) the circuit used to find the equivalent single resistance R_{AB}

of the circuit and building these up until the required equivalent resistance of the whole of the circuit is obtained. So in (c), starting at points C and D, the circuit to the left of C and D consists of a series pair of resistances (R_1 and R_2) in parallel with R_3, and the equivalent resistance can be written as

$$\frac{1}{R_{CD}} = \frac{1}{R_1 + R_2} + \frac{1}{R_3} \quad \text{or} \quad R_{CD} = \frac{(R_1 + R_2)R_3}{R_1 + R_2 + R_3}$$

Moving now to points A and B, the circuit to the left consists of a pair of series resistances (R_{CD} and R_4) in parallel with R_5. The equivalent circuit resistance R_{AB} can thus be written as

$$\frac{1}{R_{AB}} = \frac{1}{R_{CD} + R_4} + \frac{1}{R_5}$$

or

$$R_{AB} = \frac{(R_4 + R_{CD}) \cdot R_5}{R_4 + R_{CD} + R_5}$$

Substituting for R_{CD} using the previously derived expression, we obtain

$$R_{AB} = \frac{\left[\dfrac{(R_1 + R_2) \cdot R_3}{R_1 + R_2 + R_3} + R_4\right] \cdot R_5}{\dfrac{(R_1 + R_2) \cdot R_3}{R_1 + R_2 + R_3} + R_4 + R_5} \qquad \text{(A4.1)}$$

Defining I as the current flowing in the circuit when the measuring instrument is connected to it, we can write

$$I = \frac{E_o}{R_{AB} + R_m}$$

and the voltage measured by the meter is given by

$$E_m = \frac{R_m \cdot E_o}{R_{AB} + R_m}$$

In the absence of the measuring instrument and its resistance R_m, the voltage across AB would be the equivalent circuit voltage source whose value is E_o. The effect of measurement is therefore to reduce the voltage across AB by the ratio given by

$$\frac{E_m}{E_o} = \frac{R_m}{R_{AB} + R_m} \qquad \text{(A4.2)}$$

It is thus obvious that as R_m gets larger, the ratio E_m/E_o gets closer to unity, showing that the design strategy should be to make R_m as high as possible to minimize disturbance of the measured system. (Note that we did not calculate the value of E_o, since this is not required in quantifying the effect of R_m.)

Example 1

Suppose that the components of the circuit shown in Figure A4.12(a) have the following values:

$$R_1 = 400\Omega; \ R_2 = 600\Omega; \ R_3 = 1000\Omega; \ R_4 = 500\Omega; \ R_5 = 1000\Omega$$

The voltage across AB is measured by a voltmeter whose internal resistance is 9500Ω. What is the measurement error caused by the resistance of the measuring instrument?

Solution

Proceeding by applying Thevenin's theorem to find an equivalent circuit to that of Figure A4.12(a) of the form shown in Figure A4.12(b), we have from equation A4.1,

$$R_{AB} = \frac{\left[\dfrac{(R_1 + R_2) \cdot R_3}{R_1 + R_2 + R_3} + R_4\right] \cdot R_5}{\dfrac{(R_1 + R_2) \cdot R_3}{R_1 + R_2 + R_3} + R_4 + R_5}$$

Substituting for the given component values, we obtain

$$R_{AB} = \frac{\left[\dfrac{1000^2}{2000} + 500\right] \cdot 1000}{\dfrac{1000^2}{2000} + 500 + 1000} = \frac{1000^2}{2000} = 500\Omega$$

From equation (A4.2), we have

$$\frac{E_m}{E_o} = \frac{R_m}{R_{AB} + R_m}$$

The measurement error is given by $(E_o - E_m)$:

$$E_o - E_m = E_o\left[1 - \frac{R_m}{R_{AB} + R_m}\right]$$

Substituting in values:

$$E_o - E_m = E_o\left[1 - \frac{9500}{10\ 000}\right] = 0.95\ E_o$$

Thus the error in the measured value is 5 per cent.

A4.3 Bridge circuit loading during output voltage measurement

In the analysis of bridge circuits, such as that shown in Figure A4.13, the simplification of assuming zero current in the measuring instrument is often made. In practice, the voltage-measuring instrument loads the bridge circuit, because at least a small, finite current must flow into the measuring instrument.

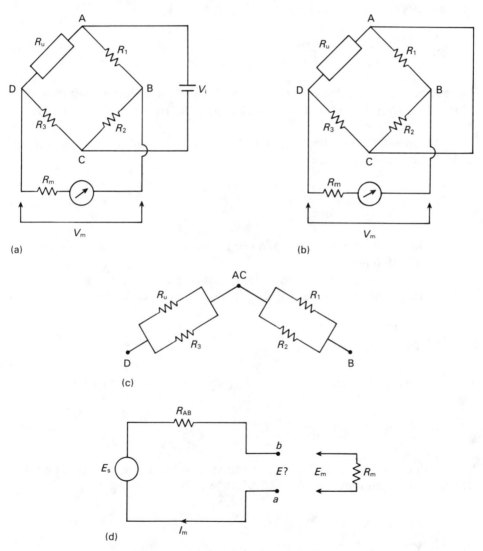

Figure A4.13: (a) Bridge circuit; (b) Bridge circuit with voltage source removed; (c) alternative representation; (d) equivalent circuit by Thevenin's theorem

The zero-current assumption clearly introduces some error which must be quantified.

Thevenin's theorem is again a useful tool for this purpose. Replacing the voltage source V_i in Figure A4.13(a) by a zero internal resistance produces the circuit shown in Figure A4.13(b) which can be shown in an equivalent representation (Figure A4.13(c)). Figure A4.13(c) shows that the equivalent circuit resistance consists of a pair of parallel resistors R_u and R_3 in series with the parallel resistor pair R_1 and R_2, and R_{DB} is given by

$$R_{DB} = \frac{R_1 \cdot R_2}{R_1 + R_2} + \frac{R_u \cdot R_3}{R_u + R_3} \tag{A4.3}$$

The equivalent circuit derived via Thevenin's theorem with the resistance, R_m, of the measuring instrument connected across the output is shown in Figure A4.13(d).

The open-circuit voltage across DB, E_o, is the output voltage for an unloaded bridge circuit ($R_m = 0$):

$$E_o = V_i \left[\frac{R_u}{R_u + R_3} - \frac{R_1}{R_1 + R_2} \right] \tag{A4.4}$$

(The derivation of equation (A4.4) can be found in many texts on measurement theory, e.g. reference 2.)

If the current flowing is I_m when the measuring instrument of resistance R_m is connected across DB, then, by Ohm's law, I_m is given by

$$I_m = \frac{E_o}{R_{DB} + R_m} \tag{A4.5}$$

If V_m is the voltage measured across R_m, then, again by Ohm's law;

$$V_m = I_m \cdot R_m = \frac{E_o \cdot R_m}{R_{DB} + R_m} \tag{A4.6}$$

Substituting for E_o and R_{DB} in equation (A4.6), using the relationships developed in equations (A4.3) and (A4.4), we obtain

$$V_m = \frac{V_i \left[\dfrac{R_u}{R_u + R_3} - \dfrac{R_1}{R_1 + R_2} \right] \cdot R_m}{\dfrac{R_1 \cdot R_2}{R_1 + R_2} + \dfrac{R_u \cdot R_3}{R_u + R_3} + R_m}$$

Simplifying:

$$V_m = \frac{V_i \cdot R_m (R_u \cdot R_2 - R_1 \cdot R_3)}{R_1 \cdot R_2 (R_u + R_3) + R_u \cdot R_3 (R_1 + R_2) + R_m (R_1 + R_2)(R_u + R_3)} \quad \text{(A4.7)}$$

Example 2

A bridge circuit as shown in Figure A4.14 is used to measure the value of the unknown resistance R_u of a strain gauge of nominal value 500Ω. The output voltage measured across points DB in the bridge is measured by a voltmeter. Calculate the measurement sensitivity in volts/ohm change in R_u if:

(a) the resistance Rm of the measuring instrument is neglected, and
(b) account is taken of the value of R_m.

Solution

For $R_u = 500Ω$, $V_m = 0$. To determine sensitivity, calculate V_m for $R_u = 501Ω$.

(a) applying equation (A4.4):

$$V_m = V_i \left[\frac{R_u}{R_u + R_3} - \frac{R_1}{R_1 + R_2} \right]$$

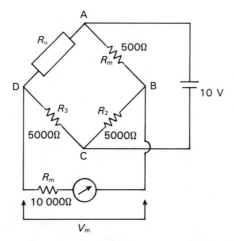

Figure A4.14 Bridge circuit – numerical example

Substituting in values:

$$V_m = \left[\frac{501}{1001} - \frac{500}{1000}\right] = 5.00 \text{ mV}$$

Thus, if the resistance of the measuring circuit is neglected, the measurement sensitivity is 5.00 mV per ohm change in R_u.

(b) applying equation (A4.7) and substituting in values:

$$V_m = \frac{10 \times 10^4 \times 500(501 - 500)}{500^2(1001) + 500 \times 501 \times 1000 + 10^4 \times 1000 \times 1001} = 4.76 \text{ mV}$$

Thus, if proper account is taken of the 10kΩ value of the resistance of R_m, the true measurement sensitivity is shown to be 4.76 mV per ohm change in R_u.

Appendix 5

Design of analogue filters

Analogue filters can be built from either active or passive elements. In the analysis below, passive filters are considered first, followed by a discussion on active filters.

Passive analogue filters

The detailed design of passive filters is a subject of some considerable complexity which is outside the scope of this book. However, it is appropriate to present the major formulae relevant to the design of these filters. These formulae are quoted without derivation and any reader requiring knowledge of the derivations is referred to one of the specialist texts referenced.[1-3]

Simple passive filters consist of a network of impedances as shown in Figure A5.1(a). So that there is no dissipation of energy in the filter, these impedances should ideally be pure reactances (capacitors or resistanceless inductors). In practice, however, it is impossible to manufacture inductors that do not have a small resistive component and so this ideal cannot be achieved. Indeed, readers familiar with radio receiver design will be aware of the existence of filters consisting of pure resistors and capacitors. These only have a mild filtering effect which is useful in radio tone controls but not relevant to the signal processing requirements discussed in this chapter. Such filters are therefore not considered further.

Each element of the network shown in Figure A5.1(a) can be represented by either a T-section or π-section, as shown in Figures A5.1(b) and A5.1(c) respectively. To obtain proper matching between filter sections, it is necessary for the input impedance of each section to be equal to the load impedance for that section. This value of impedance is known as the characteristic impedance (Z_0). For a T-section of filter, the characteristic impedance is calculated from

$$Z_0 = [Z_1 \cdot Z_2 \{1 + (Z_1/4Z_2)\}]^{1/2} \tag{A5.1}$$

The frequency attenuation characteristics of the filter can be determined by inspecting this expression for Z_0. Frequency values for which Z_0 is real lie in the pass-band and frequencies for which Z_0 is imaginary lie in the stop-band. Various types of filter can be synthesized, as shown in Figure A5.2, by appropriate choice of the components making up Z_1 and Z_2.

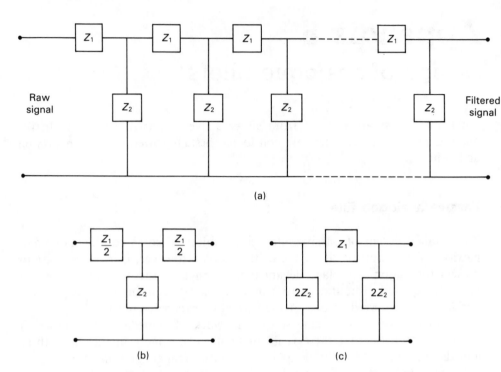

Figure A5.1: (a) simple passive filter; (b) T-section; (c) π-section

Consider the case where $Z_1 = j\omega L$ and $Z_2 = 1/j\omega C$. Substituting these values into the expression for Z_o above, we obtain

$$Z_o = [(L/C) \cdot \{1.0 - 0.25\omega^2 LC\}]^{1/2}$$

For frequencies where $\omega < \sqrt{(4/LC)}$, Z_o is real, and for higher frequencies, Z_o is imaginary. These values of impedance therefore give a *low-pass filter* with cut off frequency given by

$$f_c = (w_c/2\pi) = \frac{1}{\pi [LC]^{1/2}}$$

(NB: parameters L and C are defined with respect to the filter components expressed in Figure A5.2.)

A *high-pass filter* can be synthesized with exactly the same cut-off frequency if the impedance values chosen are

$$Z_1 = 1/j\omega C \text{ and } Z_2 = j\omega L$$

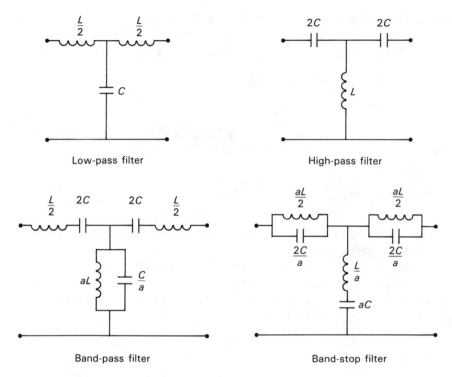

Figure A5.2 Circuit components for passive filter T-sections

(NB: parameters L and C are defined with respect to the filter components expressed in Figure A5.2.)

A point worthy of note in both these last two examples is that the product $Z_1 \cdot Z_2$ could be represented by a constant k which is independent of frequency. Because of this, such filters are known as *constant-k filters*.

A constant-k *band-pass filter* can be realized with the following choice of impedance values:

$$Z_1 = j\omega L + \frac{1}{j\omega C} \quad ; \quad Z_2 = \frac{(j\omega La)(a/j\omega C)}{j\omega La + (a/j\omega C)}$$

The frequencies f_1 and f_2 defining the end of the pass-band are most easily expressed in terms of a frequency f_0 in the centre of the pass-band. The corresponding equations are

$$f_0 = \frac{1}{2\pi(LC)^{1/2}} \quad ; \quad f_1 = f_0[(1 + a)^{1/2} - a^{1/2}] \quad ; \quad f_2 = f_0[(1 + a)^{1/2} + a^{1/2}]$$

(NB: parameters a, L and C are defined with respect to the filter components expressed in Figure A5.2.)

For a constant-k *band-stop filter*, the appropriate impedance values are

$$Z_1 = \frac{(j\omega La)(a/j\omega C)}{j\omega La + (a/j\omega C)} \quad ; \quad Z_2 = \frac{1}{a}\left(j\omega L + \frac{1}{j\omega C}\right)$$

The frequencies defining the ends of the stop-band are again normally defined in terms of the frequency f_0 in the centre of the stop-band:

$$f_0 = \frac{1}{2\pi(LC)^{1/2}} \quad ; \quad f_1 = f_0\left[1 - \frac{a}{4}\right] \quad ; \quad f_2 = f_0\left[1 + \frac{a}{4}\right]$$

(NB: parameters a, L and C are defined with respect to the filter components expressed in Figure A5.2.)

As has already been mentioned, a practical filter does not eliminate frequencies in the stop-band but merely attenuates them by a certain amount. The attenuation, α, at a frequency in the stop-band, f, for a single T-section of a low-pass filter is given by

$$\alpha = 2 \cosh^{-1}\left(\frac{f}{f_c}\right) \tag{A5.2}$$

The relatively poor attenuation characteristics are obvious if we evaluate this expression for a value of frequency close to the cut-off frequency given by $f = 2f_c$. Then $\alpha = 2 \cosh^{-1}(2) = 2.64$. Further away from the cut-off frequency, for $f = 20f_c$, $\alpha = 2 \cosh^{-1}(20) = 7.38$.

Improved attenuation characteristics can be obtained by putting several T-sections in cascade. If perfect matching is assumed then two T-sections give twice the attenuation of one section, i.e. at frequencies of $2f_c$ and $20f_c$, α for two sections would have a value of 5.28 and 14.76 respectively.

The discussion so far has assumed resistanceless inductances and perfect matching between sections. Such conditions cannot be achieved in practice and this has several consequences.

Inspection of the expression for the characteristic impedance (A5.1) reveals frequency-dependent terms. Thus the condition that the load impedance is equal to the input impedance for a section is only satisfied at one particular frequency. The load impedance is normally chosen so that this frequency is 'well within the pass-band'. (Zero frequency is usually chosen for a low-pass filter and infinite frequency for a high-pass filter as the frequency where this is satisfied.) This is one of the reasons for the degree of attenuation in the pass-band shown in the practical filter characteristics of Figure 6.7, the other reason being the presence of resistive components in the inductors of the filter. The effect of this in a practical filter is that the value of α at the cut-off frequency is 1.414 whereas the value predicted theoretically for an ideal filter (equation (A5.2)) is zero. Cascading filter sections together increases this attenuation in the pass-band as well as increasing attenuation of frequencies in the stop-band.

This problem of matching successive sections in a cascaded filter seriously degrades the performance of constant-k filters and has resulted in the development of other types such as *m*-derived and *n*-derived composite filters. These produce less attenuation within the pass-band and greater attenuation outside it than constant-k filters, although this is only achieved at the expense of greater filter complexity and cost. The reader interested in further consideration of these is directed to more specialized texts.[1-3]

Active analogue filters

In the foregoing discussion on passive filters, two main difficulties were encountered: obtaining resistanceless inductors, and achieving proper matching

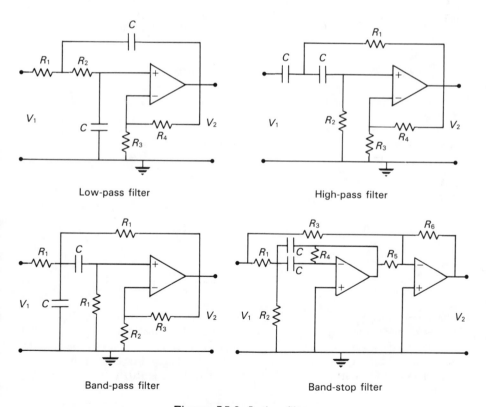

Low-pass filter   High-pass filter

Band-pass filter   Band-stop filter

Figure A5.3 Active filters

between signal source and load through the filter sections. Active filters overcome both these problems and so are very popular for signal processing duties.

Active circuits to produce the four different types of filtering identified are illustrated in Figure A5.3. These particular circuits are all second order filters because the relationship between filter input and output is described by a second order differential equation. The major component in an active filter is an electronic amplifier, with the filter characteristics being defined by a network of amplifier input and feedback components consisting of resistors and capacitors. The fact that resistors are used instead of the more expensive and bulky inductors required by passive filters is a further advantage of the active type of filter.

(a) *low-pass filter*	(b) *high-pass filter*

$$K = \frac{\mu}{R_1 R_2 C C_1}$$

$$a = \frac{1}{R_2 C_1}(1 - \mu) + \frac{1}{R_1 C} + \frac{1}{R_2 C}$$

$$b = \frac{1}{R_1 R_2 C C_1}$$

where $\mu = 1 + (R_4/R_3)$ and dc gain is K/b

$$K = \mu$$

$$a = \frac{1}{R_1 C}(1 - \mu) + \frac{2}{R_2 C}$$

$$b = \frac{1}{R_1 R_2 C^2}$$

where $\mu = 1 + (R_4/R_3)$

(c) *band-pass filter*	(d) *band-stop filter*

$$K = \frac{\mu}{R_1 C}$$

$$B = \omega_2 - \omega_1 = \frac{4 - \mu}{R_1 C}$$

$$\omega_o^2 = \frac{2}{R_1^2 C^2}$$

where $\mu = 1 + R_3/R_2$
ω_1 and ω_2 are frequencies at ends of the pass-band,
ω_o is centre frequency of the pass-band and Gain at frequency ω_o is K/b

$$R_3 R_4 = 2 R_1 R_5$$

$$B = \frac{2}{R_4 C}$$

$$\omega_o^2 = \frac{1}{R_4 C^2}\left(\frac{1}{R_1} + \frac{1}{R_2}\right)$$

where ω_1 and ω_2 are frequencies at ends of the stop-band,
ω_o is centre frequency of the pass-band and inverting gain magnitude is R_6/R_3

As in the case of passive filters, the design of active filters is a subject of considerable complexity and is only considered here in simple terms. The reader requiring a deeper understanding is referred to reference 4.

The characteristics of the filter in terms of its attenuation behaviour in the pass- and stop-bands is determined by the choice of circuit components in Figure A5.3. A common set of design formulae is given on p. 302.

Design procedures using the formulae above involve the use of graphs in which gains and cut-off frequencies are plotted for a range of parameter values (see reference 4). Suitable parameter values are then chosen by inspection.

While the above formulae represent the form of active filter which is the most general-purpose one available, many other forms also exist with the same circuit structure but with different rules for defining component values. Butterworth filters, for instance, optimize the pass-band attenuation characteristics at the expense of stop-band performance. Another form, Chebyshev filters, have very good stop-band attenuation characteristics but poorer pass-band performance. These are considered in considerable detail in reference 4.

Appendix 6
Curve fitting by regression techniques

Introduction to regression techniques

Regression techniques consist of finding a mathematical relationship between measurements of two variables y and x, such that the value of one variable y can be predicted from a measurement of the other variable x. This procedure is simplest if a straight-line relationship exists between the variables which can be estimated by linear least-squares regression.

In many cases, inspection of the raw data points plotted on a graph shows that a straight-line relationship between the points does not exist. However, knowledge of physical laws governing the data can often suggest a suitable alternative form of relationship between the two sets of variable measurements.

In some cases, the measured variables can be transformed such that a linear relationship is obtained. For example, suppose that two variables y and x are related according to

$$y = a \cdot x^c$$

A linear relationship from this can be derived as

$$\log(y) = \log(a) + c \cdot \log(x)$$

Thus if a graph is constructed of $\log(y)$ plotted against $\log(x)$, the parameters of a straight-line relationship can be estimated by linear least squares regression.

Apart from linear relationships, physical laws often suggest a relationship where one variable y is related to another variable x by a power series of the form

$$y = a_0 + a_1 \cdot x + a_2 \cdot x^2 + \dots + a_p \cdot x^p$$

Estimation of the parameters $a_0, \dots, a_p$ is very difficult if p has a large value. Fortunately, a relationship where p only has a small value can be fitted to most data sets. Quadratic least-squares regression is used to estimate parameters where p has a value of 2, and for larger values of p polynomial least-squares regression is used for parameter estimation.

Where the appropriate form of relationship between variables in measure-

ment data sets is not obvious either from visual inspection or from consideration of physical laws, a technique which is effectively a trial and error method
has to be applied. This consists of estimating the parameters of successively
higher-order relationships between y and x until a curve is found which fits the
data sufficiently closely. What level of closeness is acceptable is considered in
the later section on confidence tests.

Linear least-squares regression

If a linear relationship between y and x exists for a set of n measurements
$y_1, ..., y_n, x_1, ..., x_n$, then this relationship can be expressed as $y = a + b \cdot x$,
where the coefficients a and b are constants. The purpose of least-squares
regression is to select the optimum values for a and b such that the line gives
a good fit to the measurement data. The deviation of each point (x_i, y_i) from
the line can be expressed as d_i, where $d_i = y_i - (a + b \cdot x_i)$.

The best-fit line is obtained when the sum of the squared deviations, S, is
a minimum, i.e. when

$$S = \sum_{i=1}^{n} [d_i^2] = \sum_{i=1}^{n} [y_i - a - b \cdot x_i]^2$$

is a minimum. The minimum can be found by setting the partial derivatives $\delta S / \delta a$
and $\delta S / \delta b$ to zero and solving the resulting two simultaneous (normal) equations:

$$\delta S / \delta a = \sum 2[y_i - a - b \cdot x_i][-1] = 0$$

$$\delta S / \delta b = \sum 2[y_i - a - b \cdot x_i][-x_i] = 0$$

Using x_m and y_m to represent the mean values of x and y, the least squares
estimates $\hat{a}$ and $\hat{b}$ of the coefficients a and b can be written as

$$\hat{b} = \frac{\sum (x_i - x_m)(y_i - y_m)}{\sum (x_i - x_m)^2} \tag{A6.1}$$

$$\hat{a} = y_m - \hat{b} \cdot x_m \tag{A6.2}$$

Example
In an experiment to determine the variation of the specific heat of a

substance with temperature, the following table of results was obtained:

Temperature ($^\circ$C)	40	45	50	55	60	65	70	75	80	85	90	95	100
Specific heat		1.38	1.43	1.46	1.49	1.56	1.57	1.59	1.64	1.69	1.72	1.78	1.83 1.85

Fit a straight line to this data set using least-squares regression and estimate the specific heat at a temperature of 72 $^\circ$C.

Solution

Let y represent the specific heat and x represent the temperature. Then a suitable straight line is given by $y = a + b \cdot x$. We can now proceed to calculate estimates for the coefficients a and b using equations (A6.1) and (A6.2) above.

The first step is to calculate the mean values of x and y. These are found to be $x_m = 70$ and $y = 1.6146$. Next, we need to tabulate $(x_i - x_m)$ and $(y_i - y_m)$ for each pair of data values:

i	$(x_i - x_m)$	$(y_i - y_m)$
1	-30	-0.2346
2	-25	-0.1846
$\vdots$		
13	$+30$	$+0.2354$

Now, applying equations (A6.1) and (A6.2), we obtain:

$$\hat{b} = 0.007\ 817\ 7 \quad \text{and} \quad \hat{a} = 1.066\ 923$$

i.e.

$$y = 1.066\ 923 + 0.007\ 817\ 7x$$

Using this equation, at a temperature (x) of 72 $^\circ$C, the specific heat (y) is 1.63. Note that in this solution, we have specified the answer to an accuracy of three figures only, which is the same accuracy as the measurements. Any more figures in the answer would be meaningless.

Quadratic least-squares regression

Quadratic least-squares regression is used to estimate the parameters of a relationship $y = a + b \cdot x + c \cdot x^2$ between two sets of measurements $y_1 \dots y_n$ and $x_1 \dots x_n$.

The deviation of each point (x_i, y_i) from the line can be expressed as d_i, where $d_i = y_i - (a + b \cdot x_i + c \cdot x_i^2)$. The best-fit line is obtained when the sum of the squared deviations, S, is a minimum, i.e. when

$$S = \sum_{i=1}^{n} [d_i^2] = \sum_{i=1}^{n} [y_i - a - b \cdot x_i - c \cdot x_i^2]^2$$

is a minimum. The minimum can be found by setting the partial derivatives $\delta S/\delta a$, $\delta S/\delta b$ and $\delta S/\delta c$ to zero and solving the resulting simultaneous equations, as for the linear least-squares regression case above. Standard computer programs to estimate the parameters a, b and c by numerical methods are widely available and thus a detailed solution is not presented here.

Polynomial least-squares regression

Polynomial least-squares regression is used to estimate the parameters of the pth order relationship $y = a_0 + a_1 \cdot x + a_2 \cdot x^2 + \dots + a_p \cdot x^p$ between two sets of measurements $y_1 \dots y_n$ and $x_1 \dots x_n$.

The deviation of each point (x_i, y_i) from the line can be expressed as d_i, where

$$d_i = y_i - (a_0 + a_1 \cdot x_i + a_2 \cdot x_i^2 + \dots + a_p \cdot x_i^p)$$

The best-fit line is obtained when the sum of the squared deviations given by

$$S = \sum_{i=1}^{n} [d_i^2]$$

is a minimum. The minimum can be found as before by setting the p partial derivatives $\delta S/\delta a_0 \dots \delta S/\delta a_p$ to zero and solving the resulting simultaneous equations. Again, as for the quadratic least-squares regression case, standard computer programs to estimate the parameters $a_0 \dots a_p$ by numerical methods are widely available and thus a detailed solution is not presented here.

Confidence tests in curve fitting

Having applied least-squares regression to estimate the parameters of a chosen relationship, some form of follow-up procedure is clearly required to assess how well the estimated relationship fits the data points. One fundamental requirement in curve fitting is that the maximum deviation d_i of any data point

(y_i, x_i) from the fitted curve is less than the calculated maximum measurement error level. For some data sets, it is impossible to find a relationship between the data points which satisfies this requirement. This normally occurs when both variables in a measurement data set are subject to random variation, such as where the two sets of data values are measurements of human height and weight. Correlation analysis is applied in such cases to determine the degree of association between the variables.

Assuming that a curve can be fitted to the data points in a measurement set without violating the above fundamental requirement, a further simple curve-fitting confidence test is to calculate the sum of squared deviations S for the chosen y/x relationship and compare it with the value of S calculated for the next higher-order regression line which could be fitted to the data. Thus if a straight-line relationship is chosen, the value of S calculated should be of a similar magnitude to that obtained by fitting a quadratic relationship. If the value of S were substantially lower for a quadratic relationship, this would indicate that a quadratic relationship was a better fit to the data than a straight-line one and further tests would be needed to examine whether a cubic or higher-order relationship was a better fit still.

Other more sophisticated confidence tests exist such as the F-ratio test. However, these are outside the scope of this book.

Correlation tests

Where both variables in a measurement data set are subject to random fluctuations, correlation analysis is applied to determine the degree of association between the variables. For example, in the case already quoted of a data set containing measurements of human height and weight, we certainly expect some relationship between the variables of height and weight because a tall person is heavier *on average* than a short person. Correlation tests determine the strength of the relationship (or interdependence) between the measured variables, which is expressed in the form of a correlation coefficient.

For two sets of measurements $x_1 \dots x_n$ and $y_1 \dots y_n$ with means x_m and y_m, the correlation coefficient φ is given by

$$\varphi = \frac{\Sigma(x_i - x_m)(y_i - y_m)}{\{[\Sigma(x_i - x_m)^2][\Sigma(y_i - y_m)^2]\}^{1/2}}$$

The value of $|\varphi|$ always lies between 0 and 1, with 0 representing the case where the variables are completely independent of one another and 1 the case where they are totally related to one another.

For $0 < |\varphi| < 1$, linear least-squares regression can be applied to find

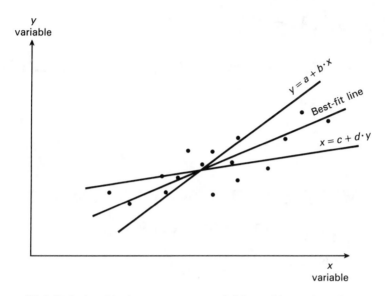

Figure A6.1 Relationship between two variables with random fluctuations

relationships between the variables which allow x to be predicted from a measurement of y and y to be predicted from a measurement of x. This involves finding two separate regression lines of the form:

$$y = a + b \cdot x \qquad \text{(regression of } y \text{ upon } x\text{)}$$

and

$$x = c + d \cdot y \qquad \text{(regression of } x \text{ upon } y\text{)}$$

These two lines are not normally coincident, as shown in Figure A6.1.

Both lines pass through the centroid of the data points but their slopes are different. As $|\varphi| \to 1$, the lines tend to coincidence, representing the case where the two variables are totally dependent upon one another. As $|\varphi| \to 0$, the lines tend to orthogonal ones parallel to the x and y axes. In this case, the two sets of variables are totally independent. The best estimate of x given any measurement of y is $\bar{x}$ and the best estimate of y given any measurement of x is $\bar{y}$.

For the general case, the best fit to the data is that line which bisects the angle between the lines on Figure A6.1.

Appendix 7

Typical structure of a quality manual

MORRIS FABRICATIONS LIMITED

QUALITY ASSURANCE MANUAL

MEASUREMENT AND CALIBRATION SYSTEMS

```
* * * * * * * * * * * * * * * * * * * * * * * * * * *
*                                                   *
*        THIS MANUAL IS A CONFIDENTIAL              *
*                 DOCUMENT                          *
*                                                   *
*      IT MUST BE KEPT SECURE AND MUST              *
*      NOT BE DISCLOSED TO NON-COMPANY              *
*      PERSONNEL EXCEPT TO BONA-FIDE                *
*         CUSTOMERS AS NECESSARY                    *
*                                                   *
* * * * * * * * * * * * * * * * * * * * * * * * * * *
```

COPY NUMBER : 4

AUTHORIZED HOLDER :

Works manager — West Cross Site

Revision number: 06   Date: 08 Sep 1989

CONFIDENTIAL

CONTENTS

Section number

Revision number: 06   Date: 08 Sep 1989

REVISION NUMBERS OF MANUAL PAGES

Page number	Revision number	Date
front page	06	08 Sep 1989
contents page	06	08 Sep 1989
1.01	01	02 Jul 1986
1.02	03	17 Nov 1986
2.01	02	09 Sep 1986
2.02	02	09 Sep 1986
2.03	06	08 Sep 1989
2.04	04	23 Feb 1988
3.01	05	31 Mar 1989
3.02	04	23 Feb 1988
4.01	01	02 Jul 1986

Revision number: 06   Date: 08 Sep 1989

AUTHORIZED HOLDERS OF MANUAL

Copy number	Job title of holder
1	Quality Assurance Director
2	Works Manager — Ringwood site
3	Quality Assurance manager — Ringwood site
4	Works Manager — West Cross site
5	Quality Assurance manager — West Cross site
6	Works Manager — Fiveoaks site
7	Quality Assurance manager — Fiveoaks site

NO FURTHER COPIES OF THIS MANUAL SHALL BE
MADE UNLESS THESE ARE AUTHORIZED BY THE
QUALITY ASSURANCE DIRECTOR AND DETAILS OF
SUCH FURTHER COPIES AND THEIR AUTHORIZED
HOLDERS ARE ENTERED ON THIS PAGE

Page number: 1.01   Revision number: 01   Date: 02 Jul 1986

AMENDMENT PROCEDURE

Any suggestions for amendments to this manual, and the quality assurance procedures described therein, should be addressed to the Quality Assurance Director. These will be considered, and any positive response will be in the form of an official revision of the manual.

NO ALTERATION WHATSOEVER MUST BE MADE TO THIS MANUAL EXCEPT FOR SUCH OFFICIAL REVISIONS MADE IN ACCORDANCE WITH THE FOLLOWING PROCEDURE.

Procedure for amendment of manual

1. Amendments as necessary will be issued by the Quality Assurance Director.
2. Each amendment package issued will consist of a set of pages, together with instructions about how these are to be integrated into the existing manual. These instructions will typically require some totally new pages to be added and some existing pages to be replaced with newer versions.
3. The amendment package will include a page which gives the correct revision levels of all pages in the manual. After updating the document, the manual holder should check all pages carefully to ensure that they are all of the correct revision level. After use, this page should be filed at the front of the manual.
4. All sheets removed from the manual in accordance with these instructions must be destroyed.
5. When the amendments to the manual have been completed, the 'Amendment Receipt Slip' provided with the amendment package should be signed and dated and returned to the Quality Assurance Director.

Page number: 1.02   Revision number: 03   Date: 17 Nov 1986

MEASUREMENT SYSTEMS SUBJECT TO CONTROL

(a) Ringwood site

Measurement requirements

This should include details of all measurements on the site which must be made to satisfy quality control and assurance requirements. For each measurement to be made, the environmental conditions existing, the measurement-limit required and the type of instrument therefore specified must be described.

Instruments used

Here, a description of each type of instrument used should be given. This should include instructions about the proper way of using each instrument and give the necessary information about any environmental control or other special precautions to be taken. A method of marking on the instrument itself such instructions about proper use and special precautions required should be prescribed if practical.

The system of serial numbers used to uniquely identify each instrument must also be defined.

The name of the person responsible for each instrument must be stated.

Page number: 2.01   Revision number: 02   Date: 09 Sep 1986

(Ringwood site continued)

Calibration

The instruments used on site should be set down in tabular form here, using the following columns:

Serial number   Type   Required calibration frequency   Person responsible

Training courses

Initial training and refresher courses to be attended by personnel using instruments are specified here.

Page number: 2.02   Revision number: 02   Date: 09 Sep 1986

(b) West Cross site

Information as above.

Page number: 2.03   Revision number: 06   Date: 08 Sep 1989

(c) Fiveoaks site

Information as above.

Page number: 2.04   Revision number: 04   Date: 23 Feb 1988

CALIBRATION PROCEDURES

Standard instruments

Here, a list of the standard instrument to be used for calibrating each type of process instrument in the system should be given, in a table with the following headings:

Type of process instrument	Type of standard instrument

Where calibration is not to be carried out internally in the company, the name of the sub-contractor nominated to perform the calibration must be specified instead of defining the standard instrument to be used.

Handling of standard instruments and traceability

For each type of standard instrument to be used, the following information must be given here:

— *How it must be stored and handled.*
— *The environmental conditions under which it must be used.*
— *The traceability of the instrument back to National Reference Standards, in the form of the list of instruments in the intermediate calibration chain. The whereabouts of the documentary evidence of this traceability, in the form of calibration certificates, etc., must be given.*

Page number: 3.01   Revision number: 05   Date: 31 Mar 1989

Recording of calibration results

In this section, the format in which calibration results are to be recorded, (i) in the site instrument manual (see Section 9.7 and Figure 9.1), and (ii) on the instruments themselves (where appropriate), should be given.

The document should specify, *as a minimum*, the following information for each instrument:

— serial number
— name of person responsible for calibration
— required calibration frequency
— date of last calibration
— calibration results

Procedure following calibration

The procedure to be followed if an instrument is found to be outside calibration limits is set out here. A foolproof procedure for marking such instruments and preventing their further use until they have been corrected must be defined.

Training courses

Details of training courses and/or refresher courses to be attended by personnel involved in calibration duties should be set down here.

CALIBRATION PROCEDURE REVIEWS

Here, the procedure used to review the continued effectiveness of the calibration system in operation is described. The information on which this review is based will consist of the incidence of customer complaints about quality plus any data available from final product inspection stations prior to delivery to customers. If any breakdowns in quality control are attributed to poor measurements, the instrument calibration system in operation must be carefully examined to determine what must be changed in order to prevent future deterioration in the quality of measurements.

The frequency at which such reviews are to take place must be defined here also.

Finally in this section, the results of each review of the calibration system must be written down. Any action taken to modify the calibration system, in order to correct for deficiencies in measurements, must also be documented.

Page number: 4.01   Revision number: 01   Date: 02 Jul 1986

References

Chapter 1

1. *BS 5750: Quality Systems*, British Standards Institution, London, 1987.
2. *ISO 9000: Quality Management and Quality Assurance Standards*, International Organisation for Standards, Geneva, 1987.
3. *EN 29000: Quality Systems*, European Committee for Standardisation (CEN), Bruxelles.

Chapter 2

1. *BSI Handbook 22: Quality Assurance*, British Standards Institution, London, 1981.
2. *BS 6143: Guide to the Determination and Use of Quality Related Costs*, British Standards Institution, London, 1990.
3. Grant, E. L. and Leavenworth, R. S. *Statistical quality control*, McGraw-Hill, 1972.
4. Juran, J. M. *Quality-control handbook*, McGraw-Hill, 1974.

Chapter 3

1. *The Operation of a Company Standards Department*, British Standards Society, London, 1979.
2. *NAMAS Document B 5103: Certificates of Calibration*, NAMAS Executive, National Physical Laboratory, Middlesex, UK, 1985.
3. *ISO GUIDE 25: General Requirements for the Technical Competence of Testing Laboratories*, International Organisation for Standards, Geneva, 1982.
4. *BS 6460: Accreditation of Testing Laboratories*, British Standards Institution, London, 1983.

5. *BS 5497: Guide for the Determination of Repeatability and Reproducibility for a Standard Test Method by Inter-laboratory Tests*, British Standards Institution, London, 1987.
6. *ISO 5725: Precision of Test Methods – Determination of Repeatability and Reproducibility by Inter-laboratory Tests*, International Organisation for Standards, Geneva, 1986.

Chapter 4

1. *BS 5233: Glossary of Terms Used in Metrology* (incorporating BS 2643), British Standards Institution, London, 1986.
2. *BS 5532: Statistical Terminology*, British Standards Institution, London, 1978.
3. *ISO 3534: Statistics – Vocabulary and Symbols*, International Organisation for Standards, Geneva, 1977.

Chapter 6

1. Huelsman, L. P. *Active Filters: Lumped, distributed, integrated, digital and parametric*, McGraw-Hill, 1970.
2. Lynn, P. A. *The Analysis and Processing of Signals*, Macmillan, 1973.

Chapter 7

1. *BS 5532: Statistical Terminology*, British Standards Institution, London, 1978.
2. *ISO 3534: Statistics – Vocabulary and Symbols*, International Organisation for Standards, Geneva, 1977.

Chapter 8

1. Dhillon, B. S. and Singh, C. *Engineering Reliability: New techniques and applications*, Wiley, 1981.

Chapter 9

1. Morris, A. S. *Principles of measurement and instrumentation*, Prentice-Hall, 1988.

Chapter 10

1. *NAMAS Document B 5551: Thermocouples, Reference Tables and Traceability*, NAMAS Executive, National Physical Laboratory, Middlesex, UK, 1975.

2. Morris, A. S. *Principles of Measurement and Instrumentation*, Prentice-Hall, 1988.
3. Moore, G. 'Acoustic thermometry', *Electronics and Power*, **30**, 675–7, 1984.
4. *NAMAS Document NIS 7: Traceability – Thermometers, Thermocouples and Platinum Resistance Thermometers*, NAMAS Executive, National Physical Laboratory, Middlesex, UK, 1984.

Chapter 11

1. Benedict, R. P. *Fundamentals of Temperature, Pressure and Flow Measurement*, Wiley, 1984.

Chapter 12

1. *NAMAS Document NIS 6: Traceability – Weighing Equipment and Weight*, NAMAS Executive, National Physical Laboratory, Middlesex, UK, 1984.

Chapter 13

1. Anthony, D. M. *Engineering Metrology*, Pergamon, 1986.
2. Hume, K. J. *Engineering Metrology*, McDonald, 1970.
3. *BS 817: Specifications for Surface Plates*, British Standards Institution, London, 1988.
4. *BS 4372: Specifications for Engineers' Steel Measuring Rules*, British Standards Institution, London, 1968.
5. *ISO 8322: Building Construction – Measuring Instruments – Procedures for Determining Accuracy in Use*, Part 2 – *Measuring Tapes*, International Organisation for Standards, Geneva, 1989.
6. *BS 4035: Specifications for Linear Measuring Instruments for Use in Building and Civil Engineering Constructional Works. Steel Measuring Tapes, Steel Bands and Retractable Steel Pocket Rules*, British Standards Institution, London, 1966.
7. *BS 887: Specifications for Precision Vernier Calipers*, British Standards Institution, London, 1982.
8. *ISO 6906: Vernier Calipers Reading to 0.02 mm.*, International Organisation for Standards, Geneva, 1984.
9. *BS 870: Specifications for External Micrometers*, British Standards Institution, London, 1950.
10. *ISO 3611: Micrometer Calipers for External Measurement*, International Organisation for Standards, Geneva, 1978.
11. *BS 959: Specifications for Internal Micrometers* (including stick micrometers), British Standards Institution, London, 1950.
12. *BS 4311 (Parts 1 and 2): Gauge Blocks and Accessories*, British Standards Institution, London, 1968.

13. *ISO 3650: Gauge Blocks*, International Organisation for Standards, Geneva, 1978.
14. *BS 5317: Specification of Metric Length Bars and their Accessories*, British Standards Institution, London, 1976.
15. *BS 1643: Specifications for Precision Vernier Height Gauges*, British Standards Institution, London, 1983.
16. *BS 907: Specifications for Dial Gauges for Linear Measurements*, British Standards Institution, London, 1965.
17. *ISO/R 463: Dial Gauges Reading in 0.01 mm., 0.001 inch and 0.0001 inch*, International Organisation for Standards, Geneva, 1965.
18. *BS 1685: Specifications for Bevel protractors* (mechanical and optical), British Standards Institution, London, 1951.
19. *BS 958: Specifications for Spirit Levels for Use in Precision Engineering*, British Standards Institution, London, 1968.
20. *NAMAS Document B 7002: Length Bars − Calibration Procedures and Uncertainties*, NAMAS Executive, National Physical Laboratory, Middlesex, UK, 1982.
21. *BS 3064: Specifications for Metric Sine Bars and Sine Tables* (excluding compound tables), British Standards Institution, London, 1978.

Chapter 14

1. Bentley, J. P. *Principles of Measurement Systems*, Longman, 1975.
2. Britton, C. and Mesnard, D. 'A performance summary of averaging pitot-type primaries', *Measurement and Control*, **15**, 341−50, 1982.
3. King, N. W. 'Multi-phase flow measurement at NEL', *Measurement and Control*, **21**, 237−39, 1988.
4. Blickley, G. J. 'Magmeters require less power', *Control Engineering*, August, p. 65, 1987.
5. Kinghorn, F. C. 'Challenging areas in flow measurement', *Measurement and Control*, **21**, 229−35, 1988.
6. Medlock, R. S. 'Cross-correlation flow measurement', *Measurement and Control*, **18**(8), 293−8, 1985.
7. Berman, M. and Prowse, D. B. 'The statistical properties of the number of observations used for proving meters', *Measurement and Control*, **19**, 158−60, 1986.
8. Pursley, W. C. 'The calibration of flowmeters' *Measurement and Control*, **19**(5), 37−45, 1986.

Chapter 15

1. *The Instrument Manual*, Ed. J. T. Miller, United Trade Press, pp. 62−106, 1975.
2. Cheng, D. C-H, 'Viscosity: calibration and standards', *Measurement and Control*, **14**, 73−8, 1981.

3. *BS 5497: Guide for the Determination of Repeatability and Reproducibility for a Standard Test Method by Inter-laboratory Tests*, British Standards Institution, London, 1987.
4. *ISO 5725: Precision of Test Methods – Determination of Repeatability and Reproducibility by Inter-laboratory Tests*, International Organisation for Standards, Geneva, 1986.
5. Anderson, J. G. 'Paper moisture measurement using microwaves', *Measurement and Control*, **22**, 82–4, 1989.
6. Thompson, F. 'Moisture measurement using microwaves', Ibid, pp. 210–15.
7. Slight, H. A. 'Further thoughts on moisture measurement', Ibid, pp. 85–6.
8. Gimson, C. 'Using the capacitance charge transfer principle for water content measurement', Ibid, pp. 79–81.
9. Young, L. 'Moisture measurement using low resolution nuclear magnetic resonance', Ibid, pp. 54–5.
10. Benson, I. B. 'Industrial applications of near infrared reflectance for the measurement of moisture', Ibid, 45–9.
11. Wiltshire, M. P. 'Ultrasonic moisture measurement', Ibid, pp. 51–3.
12. *The Instrument Manual*, Ed. J. T. Miller, United Trade Press, pp. 180–209, 1975.
13. Pragnell, R. F. 'The modern condensation dewpoint hygrometer', *Measurement and Control*, **22**, 74–7, 1989.
14. *NAMAS Document NIS 5: Traceability – Volumetric Glassware*, NAMAS Executive, National Physical Laboratory, Middlesex, UK, 1984.

Appendix 4

1. Skilling, H. H. *Electrical Engineering Circuits*, Wiley, 1967.
2. Morris, A. S. *Principles of Measurement and Instrumentation*, Prentice-Hall, 1988.

Appendix 5

1. Blinchikoff, H. J. *Filtering in the Time and Frequency Domains*, Wiley, 1976.
2. Skilling, H. H. *Electrical Engineering Circuits*, Wiley, 1967.
3. Williams, E. *Electric Filter Circuits*, Pitman, 1963.
4. Hilburn, J. L. and Johnson, D. E. *Manual of Active Filter Design*, McGraw-Hill, 1973.

Index